VEHICULAR 2023

The Twelfth International Conference on Advances in Vehicular Systems, Technologies and Applications

March 13th - 17th, 2023

Barcelona, Spain

VEHICULAR 2023 Editors

Reiner Kriesten, Karlsruhe University of Applied Sciences, Germany

Panos Nasiopoulos, University of British Columbia, Canada

Printed from e-media with permission by:

Curran Associates, Inc.
57 Morehouse Lane
Red Hook, NY 12571

Some format issues inherent in the e-media version may also appear in this print version.

International Academy, Research, and Industry Association (IARIA)
412 Derby Way
Wilmington, DE 19810

Phone: (408) 893-6407
Fax: (408) 527-6351

petre@iaria.org

Additional copies of this publication are available from:

Curran Associates, Inc.
57 Morehouse Lane
Red Hook, NY 12571 USA
Phone: 845-758-0400
Fax: 845-758-2633
Email: curran@proceedings.com
Web: www.proceedings.com

Table of Contents

Intentional Binding in Direct and Indirect Manipulation for Pointing Task 1
Kazuhisa Miwa, Mizuki Hirata, Hojun Choi, and Tomomi Shimizu

A Neural Network-Based Estimation of Tire Self-Aligning Torque 5
Younesse El Mrhasli, Bruno Monsuez, and Xavier Mouton

Towards Cross-Cultural Intelligent Vehicles: A Review 12
Imane Taourarti, Arunkumar Ramaswamy, Javier Ibanez-guzman, Bruno Monsuez, and Adriana Tapus

Federated Monocular 3D Object Detection for Autonomous Driving 20
Fangyuan Chi, Yixiao Wang, Panos Nasiopoulos, Victor C.M. Leung, and Mahsa T. Pourazad

Simulation Environment for Validation of Automated Lane-Keeping System 24
Jiri Vlasak, Michal Sojka, and Zdene?k Hanza?lek

Baseline Selection for Integrated Gradients in Predictive Maintenance of Volvo Trucks' Turbocharger 29
Nellie Karlsson, My Bengtsson, Mahmoud Rahat, and Peyman Sheikholharam Mashhadi

An Efficient YOLOv7x Based Automated Street Parking Space Detection for Smart Cities 37
Tala Bazzaza, Hamid Reza Tohidypour, Yixiao Wang, and Panos Nasiopoulos

Approach to Modeling Energy Consumption of Auxiliary Consumers in Electric Vehicles 41
Tuyen Nguyen, Hendrik Schulte, and Reiner Kriesten

A Holistic Approach on Automotive Cybersecurity for Suppliers 47
Jose Angel Gumiel Quintana, Jon Mabe Alvarez, Jaime Jimenez Verde, and Jon Barruetabena Pujana

VEHICULAR 2023

Forward

The Twelfth International Conference on Advances in Vehicular Systems, Technologies and Applications (VEHICULAR 2023), held between March 13th and March 17th, 2023, continued a series of events considering the state-of-the-art technologies for information dissemination in vehicle-to-vehicle and vehicle-to-infrastructure and focusing on advances in vehicular systems, technologies and applications.

Mobility brought new dimensions to communication and networking systems, making possible new applications and services in vehicular systems. Wireless networking and communication between vehicles and with the infrastructure have specific characteristics from other conventional wireless networking systems and applications (rapidly changing topology, specific road direction of vehicle movements, etc.). These led to specific constraints and optimizations techniques; for example, power efficiency is not as important for vehicle communications as it is for traditional ad hoc networking. Additionally, vehicle applications demand strict communications performance requirements that are not present in conventional wireless networks. Services can range from time-critical safety services, traffic management, to infotainment and local advertising services. They are introducing critical and subliminal information. Subliminally delivered information, unobtrusive techniques for driver's state detection, and mitigation or regulation interfaces enlarge the spectrum of challenges in vehicular systems.

We take here the opportunity to warmly thank all the members of the VEHICULAR 2023 technical program committee, as well as all the reviewers. The creation of such a high-quality conference program would not have been possible without their involvement. We also kindly thank all the authors who dedicated much of their time and effort to contribute to VEHICULAR 2023. We truly believe that, thanks to all these efforts, the final conference program consisted of top-quality contributions. We also thank the members of the VEHICULAR 2023 organizing committee for their help in handling the logistics of this event.

We hope that VEHICULAR 2023 was a successful international forum for the exchange of ideas and results between academia and industry and for the promotion of progress in the field of vehicular systems, technologies and applications.

VEHICULAR 2023 Chairs

VEHICULAR 2023 Steering Committee

Khalil El-Khatib, University of Ontario Institute of Technology – Oshawa, Canada
Éric Renault, ESIEE Paris, France
Xiaohong Peng, Birmingham City University, UK
Yuping He, University of Ontario Institute of Technology, Canada

VEHICULAR 2023 Publicity Chairs

José Miguel Jiménez, Universitat Politecnica de Valencia, Spain
Sandra Viciano Tudela, Universitat Politecnica de Valencia, Spain

VEHICULAR 2023
Committee

VEHICULAR 2023 Steering Committee

Khalil El-Khatib, University of Ontario Institute of Technology – Oshawa, Canada
Éric Renault, ESIEE Paris, France
Xiaohong Peng, Birmingham City University, UK
Yuping He, University of Ontario Institute of Technology, Canada

VEHICULAR 2023 Publicity Chairs

José Miguel Jiménez, Universitat Politecnica de Valencia, Spain
Sandra Viciano Tudela, Universitat Politecnica de Valencia, Spain

VEHICULAR 2023 Technical Program Committee

Vahdat Abdelzad, University of Waterloo, Canada
Nor Fadzilah Abdullah, National University of Malaysia (UKM), Malaysia
Mohammed Al-Ansi, University Malaysia Perlis (UniMAP), Malaysia
Sufyan T. Faraj Al-Janabi, University of Anbar, Ramadi, Iraq
Mustafa S. Al-Jumaily, University of Tennessee, Knoxville, USA
Ali Alfoudi, Al-Qadisiyah University, Irak
Ala'a Al-Momani, Ulm University, Germany
Bhaskar Anand, Indian Institute of Technology Hyderabad, India
Adel Aneiba, Birmingham City University, UK
Muhammad Asim, Chung-Ang University, Seoul, South Korea
Hakan Aydin, Karadeniz Technical University, Turkey
Manlio Bacco, CNR-ISTI, Italy
Andrea Baiocchi, University of Roma "Sapienza", Italy
Paolo Barsocchi, ISTI (Institute of Information Science and Technologies) | Italian National Research Council (C.N.R.), Pisa, Italy
Paulo C. Bartolomeu, University of Aveiro, Portugal
Marcel Baunach, Graz University of Technology, Austria
Rahim (Ray) Benekohal, University of Illinois at Urbana-Champaign, USA
Sylvia Bhattacharya, Kennesaw State University, USA
Konstantinos Blekas, University of Ioannina, Greece
Eugen Borcoci, University Politehnica of Bucharest, Romania
Alexandros-Apostolos A. Boulogeorgos, University of Piraeus, Greece
Christos Bouras, University of Patras, Greece
Alyssa Byrnes, University of North Carolina at Chapel Hill, USA
Roberto Caldelli, CNIT, Florence, Italy
Rodrigo Capobianco Guido, São Paulo State University (UNESP), Brazil
Antonio Carcaterra, Sapienza University of Rome, Italy
Pedro Cardoso, Universidade do Algarve, Portugal
Marcos F. Caetano, University of Brasilia, Brazil
Florent Carlier, Centre de Reserche en Education de Nantes / Le Mans Université, France

Juan Carlos Ruiz, Universitat Politecnica de Valencia, Spain
Rodrigo Castelan Carlson, Federal University of Santa Catarina, Brazil
Francois Chan, Royal Military College, Canada
Claude Chaudet, Webster University Geneva, Switzerland
Rui Chen, Xidian University, China
Gihwan Cho, Jeonbuk University, Korea
Adam Cohen, University of California, Berkeley, USA
Gianpiero Costantino, Institute of Informatics and Telematics (IIT) | National Research Council (CNR), Italy
Yousef-Awwad Daraghmi, Palestine Technical University-Kadoorie, Palestine
David de Andrés, Universitat Politècnica de València, Spain
Fawad Ud Din, McGill University, Canada
Liza Dixon, Independent Researcher, Germany
Zoran Duric, George Mason University, USA
Péter Ekler, Budapest University of Technology and Economics, Hungary
Suzi Iryanti Fadilah, Universiti Sains Malaysia, University
Mariano Falcitelli, Photonic Networks & Technologies National Laboratory of CNIT, Italy
Abraham O. Fapojuwo, University of Calgary, Canada
Gustavo Fernandez Dominguez, Center for Digital Safety & Security | AIT Austrian Institute of Technology, Austria
Irene Fiesoli, University of Florence, Italy
Francesco Flammini, Mälardalen University, Sweden
Miguel Franklin de Castro, Federal University of Ceará, Brazil
Tomonari Furukawa, University of Virginia, USA
Varun Garg, UMass Lowell, USA
Pedro Pablo Garrido Abenza, Universidad Miguel Hernandez de Elche, Spain
Malek Ghanes, Centrale Nantes, France
Lee Gillam, University of Surrey, UK
Apostolos Gkamas, University Ecclesiastical Academy of Vella of Ioannina, Greece
Sezer Goren, Yeditepe University, Turkey
Alberto Gotta, National Research Council - Institute of Information Science and Technologies "A. Faedo" (CNR-ISTI), Italy
Heinrich Gotzig, Valeo, Germany
Javier Gozalvez, Universidad Miguel Hernandez de Elche, Spain
Kacper Grzedzinski, Cranfield University, UK
Abel Guilhermino da Silva Filho, Federal University of Pernambuco - UFPE, Brazil
Marco Häberle, University of Tuebingen, Germany
Rami Hamdi, Hamad Bin Khalifa University, Qatar Foundation, Qatar
Kyungtae (KT) Han, Toyota North America, USA
Petr Hanáček, Brno University of Technology, Czech Republic
Hong Hande, Huawei Technologies, Singapore
Yuping He, University of Ontario Institute of Technology, Canada
Gonçalo Homem de Almeida Correia, TU Delft, Netherlands
Moritz Höser, Bauhaus Luftfahrt e. V., Taufkirchen, Germany
Javier Ibanez-Guzman, Renault S.A., France
Hocine Imine, IFSTTAR/LEPSIS, France
Uzair Javaid, National University of Singapore, Singapore
Dush Nalin Jayakody, Tomsk Polytechnic University, Russia

Terje Jensen, Telenor, Norway
Yiming Ji, Georgia Southern University, USA
Felipe Jiménez Alonso, Technical University of Madrid, Spain
Magnus Jonsson, Halmstad University, Sweden
Filbert Juwono, Curtin University, Malaysia
Thabet Kacem, University of the District of Columbia, USA
Chuyo Kaku, Jiangsu Chaoli Electric Co. Ltd., China
Gorkem Kar, Bahcesehir University, Turkey
Frank Kargl, Institute of Distributed Systems | Ulm University, Germany
Sokratis K. Katsikas, Norwegian University of Science and Technology, Norway
Tetsuya Kawanishi, Waseda University, Japan
John Kenney, Toyota InfoTech Labs, USA
Norazlina Khamis, Universiti Malaysia Sabah, Malaysia
Asil Koc, McGill University, Montreal, Canada
Lisimachos Kondi, University of Ioannina, Greece
Spyros Kontogiannis, University of Ioannina, Greece
Anastasios Kouvelas, ETH Zürich, Switzerland
Zdzislaw Kowalczuk, Gdansk University of Technology, Poland
Francine Krief, Bordeaux INP, France
Reiner Kriesten, University of Applied Sciences Karlsruhe, Germany
Ryo Kurachi, Nagoya University, Japan
Gyu Myoung Lee, Liverpool John Moores University, UK
Sebastian Lindner, Hamburg University of Technology, Germany
Lianggui Liu, Zhejiang Shuren University, China
Tomasz Mach, Samsung R&D Institute, UK
Dalila B. Megherbi, University of Massachusetts, USA
Zoubir Mammeri, IRIT - Paul Sabatier University, France
Sunil Manvi, REVA University, India
Mirco Marchetti, University of Modena and Reggio Emilia, Italy
Philippe Martinet, Inria Center at Côte d'Azur University, France
P. Takis Mathiopoulos, University of Athens, Greece
Ioannis Mavromatis, University of Bristol, UK
Rashid Mehmood, King Abdul Aziz University, Saudi Arabia
José Manuel Menéndez, Universidad Politécnica de Madrid, Spain
Maria Luisa Merani, University of Modena and Reggio Emilia, Italy
Lyudmila Mihaylova, The University of Sheffield, UK
Gerardo Mino Aguilar, Benemérita Universidad Autónoma De Puebla, Mexico
Naveen Mohan, KTH, Sweden
Bruno Monsuez, ENSTA ParisTech, France
Luís Moutinho, Escola Superior de Gestão e Tecnologia de Águeda (ESTGA) - University of Aveiro / Instituto de Telecomunicações, Portugal
Vasco N. G. J. Soares, Instituto de Telecomunicações / Instituto Politécnico de Castelo Branco, Portugal
Jose Eugenio Naranjo Hernandez, Universidad Politécnica de Madrid | Instituto Universitario de Investigación del Automóvil (INSIA), Spain
Ridha Nasri, Orange Labs, France
Keivan Navaie, Lancaster University, UK
Tomas Olovsson, Chalmers University of Technology, Gothenburg, Sweden
Rodolfo Orjuela, IRIMAS-Université de Haute-Alsace (UHA), France

Patrik Österberg, Mid Sweden University, Sweden
Gerald Ostermayer, University of Applied Sciences Upper Austria, Hagenberg, Austria
Ilias Panagiotopoulos, Harokopio University of Athens, Greece
Sugandh Pargal, IIT Kharagpur, India
Antonio M. Pascoal, Institute for Systems and Robotics - IST | Univ. Lisbon, Portugal
Al-Sakib Khan Pathan, United International University, Bangladesh
Xiaohong Peng, Birmingham City University, UK
Paulo J. G. Pereirinha, Polytechnic of Coimbra | IPC/ISEC | INESC Coimbra, Portugal
Fernando Pereñíguez García, University Defense Center | Spanish Air Force Academy, Murcia, Spain
Valerio Persico, University of Napoli Federico II, Italy
Ivan Petrunin, Cranfield University, UK
Paulo Pinto, Universidade Nova de Lisboa, Portugal
Srinivas Pulugurtha, The University of North Carolina at Charlotte, USA
Hesham Rakha, Virginia Tech, USA
Mohd Fadlee A Rasid, Universiti Putra Malaysia, Malaysia
Éric Renault, ESIEE Paris, France
Martin Ring, Robert Bosch GmbH, Germany
Geraldo P. Rocha Filho, University of Brasília, Brazil
Aymeric Rousseau, Argonne National Laboratory, USA
Javier Rubio-Loyola, CINVESTAV, Mexico
João Rufino, Instituto de Telecomunicações - Pólo Aveiro, Portugal
Aduwati Sali, UPM, Malaysia
José Santa Lozano, Technical University of Cartagena (UPCT), Spain
José Santa, Technical University of Cartagena, Spain
Nico Saputro, Parahyangan Catholic University, Bandung, Indonesia
Marios Savvides, Carnegie Mellon University, USA
Erwin Schoitsch, AIT Austrian Institute of Technology GmbH, Austria
Barbara Schütt, FZI Research Center of Information Technology, Germany
Alireza Shahrabi, Glasgow Caledonian University, Scotland, UK
Rajan Shankaran, Macquarie University, Australia
Prinkle Sharma, University of Massachusetts Dartmouth, USA
Shih-Lung Shaw, University of Tennessee Knoxville, USA
Dana Simian, Lucian Blaga University of Sibiu, Romania
Pranav Kumar Singh, Central Institute of Technology Kokrajhar, India
Michal Sojka, Czech Technical University in Prague, Czech Republic
Marko Sonkki, Ericsson, Germany
Chokri Souani, Higher Institute of Applied Sciences and Technology of Sousse, Tunisia
Essam Sourour, Prince Sattam Bin Abdul-Aziz University (PSAU), Saudi Arabia
Mujdat Soyturk, Marmara University, Turkey
Anand Srivastava, IIIT Delhi, India
Pawan Subedi, The University of Alabama, Tuscaloosa, USA
Qasim Sultan, Chung-Ang University, Seoul, South Korea
Qingquan Sun, California State University San Bernardino, USA
Akimasa Suzuki, Iwate Prefectural University, Japan
Philipp Svoboda, TU Wien, Austria
Wai Yuen Szeto, The University of Hong Kong, Hong Kong
Getaneh Berie Tarekegn, National Taipei University of Technology, Taiwan
Angelo Trotta, University of Bologna, Italy

Copyright Information

Intentional Binding in Direct and Indirect Manipulation for Pointing Task

Kazuhisa Miwa
Graduate School of Informatics
Nagoya University
Nagoya 4648601, Japan
email: miwa@is.nagoya-u.ac.jp

Mizuki Hirata
Graduate School of Informatics
Nagoya University
Nagoya 4648601, Japan
email: hirata.mizuki.f1@s.mail.nagoya-u.ac.jp

Hojun Choi
Graduate School of Informatics
Nagoya University
Nagoya 4648601, Japan
email: choi.hojun.f4@s.mail.nagoya-u.ac.jp

Tomomi Shimizu
Advanced Development Div.,
Tokai Rika Co., Ltd.
Toyota, Oguchi-cho, Niwa-gun, Aichi 4800195, Japan
email: tomomi.shimizu@exc.tokai-rika.co.jp

***Abstract*—Sense Of Agency (SOA) has the potential to be a useful indicator for the evaluation of the indirect manipulation of devices in in-car systems. In this study, we examined whether intentional binding, which is used as an implicit evaluation index of SOA, is confirmed when engaging in a task in which a series of sequential operations are required. Sixteen subjects were tested, and the results showed that both direct and indirect manipulation of devices indicated the same patterns of intentional binding as in previous studies. However, no difference in the magnitude of binding was detected between direct and indirect manipulation.**

Keywords - Sense of agency; Intentional binding; Indirect manipulation.

I. INTRODUCTION

A. Indirect manipulation

In recent years, automotive infotainment systems have been developed to integrate advanced driver assistance and entertainment functions in addition to vehicle-related information. Consequently, Human Machine Interactions (HMIs) have become highly complex [1].

In this context, recent in-car devices incorporate interfaces that can be operated using touch, voice, and gestures. Early in-car devices were operated by directly touching a physical interface, consisting of combinations of knobs, buttons, and switches. However, as devices became increasingly digitalized, the target device was operated not by touching it directly, but via an input device that mediated the operation. For example, a navigation system installed on a dashboard could be operated using a touch panel. In this paper, the former is called direct operation, and the latter is called indirect operation.

In indirect manipulation, operations are performed by a cursor and a pointer that are displayed on the operating device. Such a cursor and pointer are considered extensions of the user's body (in this case, a finger), which Seinfeld et al. call user representations [2]. In an indirect operation, the space that the user touches for manipulation is called the input space (in the above example, the touch panel), and the display screen of the target device is called the output space (the navigation system operation screen).

In indirect manipulation, two interfaces, the input space and output space, arise. One of the drawbacks with the emergence of these two interfaces is the lack of feeling of direct manipulation. Lack of feeling causes problems in that, additional cognitive resources must be allocated to manipulating the device, and the user experience is degraded. It is important to clarify how this sense of direct manipulation can be maintained during the design and evaluation of increasingly complex in-car systems.

B. Sense of Agency

An important concept related to the feeling of direct manipulation is Sense Of Agency (SOA) [3]; SOA is the sense that one is the subject who causes the result that appears, and that one intentionally controls that result. In recent years, the importance of SOA has been widely recognized in the field of human-computer interaction [4].

For instance, in the field of VR (Virtual Reality), how one perceives SOA for one's own avatar in a VR space has been intensively studied [5]. Avatars are considered to be one form of user representation, as mentioned above. In indirect manipulation in in-car device operation, the feeling of SOA in the movements of a cursor or pointer displayed in the output space is an essential requirement for improving the feeling of direct manipulation.

There are two main types of SOA measurements: explicit and implicit measurements [3][6]. The most common method of explicit measurement is to directly and subjectively rate the degree to which participants perceive SOA. This method is widely used in many SOA studies because it is easy to implement, and its usefulness has been confirmed. On the other hand, it is pointed out that such subjective ratings are prone to many cognitive biases, and its limitations have been discussed [7].

C. Intentional Binding

In contrast, implicit measurement measures SOA using behavioral indicators that are not directly related to the subjective sense of SOA. The most widely used method is the one using Intentional Binding [8][9], a phenomenon in which the time interval between intentional action and the sensory stimulus caused by the action, which is fed back after a certain time is perceived as short. It has been widely confirmed that SOA is related to Intentional Binding, and a method of measuring

Intentional Binding using a temporal sensory evaluation device called the Libet clock has been established and widely used in SOA research.

On the other hand, measurements using the Libet clock requires high visual attention to the clock displayed on the screen. Therefore, there are significant limitations to its use, such as interference with the context of the main task and heavy measurement loads on participants. In recent years, new methods for measuring Intentional Binding have been developed.

Early experiments on Intentional Binding used button pressing as an intentional action and audio feedback as the result of that action. In recent years, various intentional actions have been taken, such as input by finger movement in a hollow space without a physical input device [10], and experiments using auditory and tactile stimuli have also been conducted for feedback [11][12]. Thus, examining the circumstances under which Intentional Binding occurs is important in SOA research.

D. The present study

As mentioned above, in-car devices have become increasingly digitalized in recent years, and indirect manipulation has become mainstream. In this context, a major issue is ensuring the feeling of the direct manipulation of complex in-car devices. Intentional binding is a valuable indicator for evaluating the feeling of direct manipulation.

So far, Intentional Binding measurement has been examined using simple action-feedback pairs. However, when attempting to apply this method to the development of devices for in-car systems, it is necessary to measure Intentional Binding in situations in which a series of sequential operations is engaged. Therefore, in this study, we measure Intentional Binding when engaging in such a task. We then answered the following two research questions.

- Research Question 1:
  Is Intentional Binding confirmed for each of direct manipulation and indirect manipulation?
- Research Question 2:
  Is there a difference in the degree of Intentional Binding between direct and indirect manipulations?

Section 2 introduces the summary of the experiment, and the measures of the results. Section 3 indicates the experimental results. Section 4 summarizes the discussion and conclusions.

II. EXPERIMENT

A. Participants

Sixteen participants, recruited from the general public, participated in the experiment (8 males, 8 females, age: M=46.69, SD = 16.62). All the participants were right-handed.

B. Task

Figure 1 shows a screenshot of the task in which the participants engaged. Participants were asked to trace the numbers from 1 to 3 displayed on the tablet in sequence, and finally, to tap the last number 3. Upon tapping the number 3, a beep sound was fed back approximately 250 ms later. The Libet clock was displayed so that they reported the timing of tapping and hearing the beep sound.

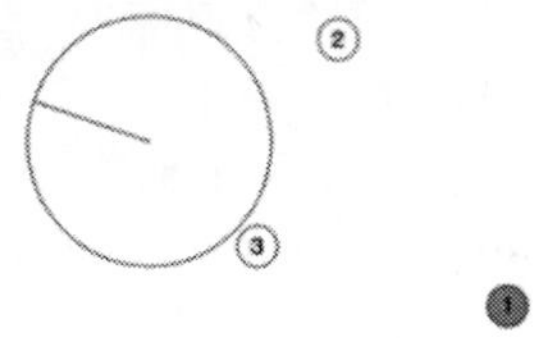

Figure 1. Screenshot of the task.

C. Experimental Design

The experimental design was a three-factors (Direct-Indirect × Action-Outcome × Baseline-Operant) within-participant factorial design.

1) Direct-Indirect Factor: In the Direct condition, participants directly manipulated the tablet (Figure 2 (a)). In the Indirect condition, participants indirectly manipulated the target tablet by tracing another tablet placed underneath the target tablet (Figure 2 (b)). In the Indirect condition, participants could not see their own hands as they manipulated.

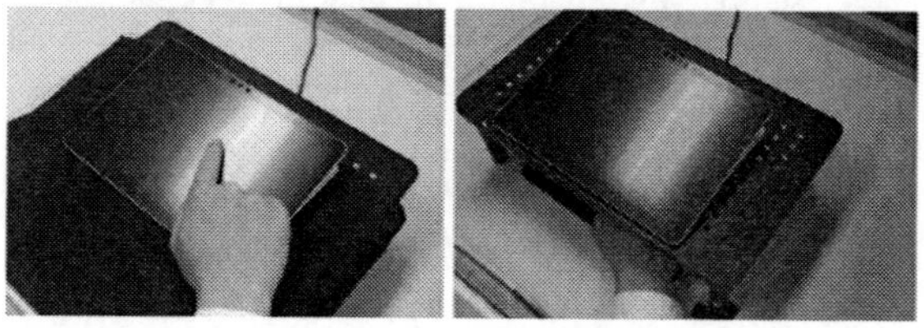

(a) Direct condition (b) Indirect condition

Figure 2. Direct and Indirect conditions of experiment.

The same tablet was used as the input and display devices in the Indirect condition, and the input from the input device was output to the display device for presentation. Therefore, the hardware performance under both conditions was identical, including the time lag between the finger movements reflected in the pointer displayed on the display tablet.

Figure 3 illustrates the settings controlled by the Action-Outcome factor and the Baseline-Operant factor.

2) Action-Outcome Factor: In the Action conditions, the participant's perceived timing when the last 3 number was tapped was measured using the Libet clock, while in the Outcome conditions, the participant's perceived timing when the beep sounded was measured.

3) Baseline-Operant Factor: In the Baseline conditions, no beep sounded in response to the participant's action; or a beep sounded without the participant's action. In the Operant conditions, a beep sounded approximately 250 ms seconds after the participant's action.

D. Procedure

The participants were informed about the overview of the experiment, followed by a practice phase. They then moved on to the main phase, which consisted of twelve trials for each condition.

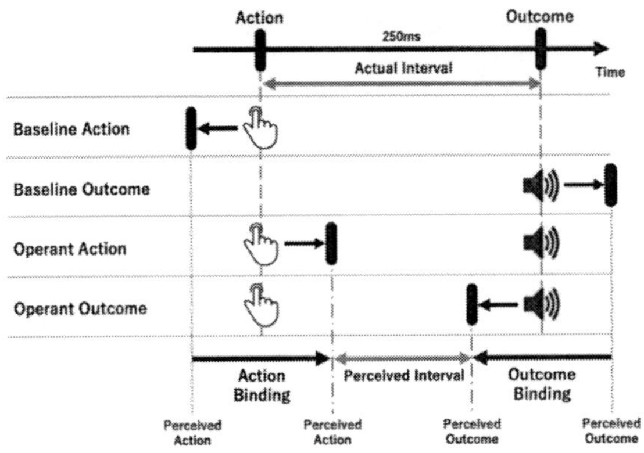

Figure 3. Action Binding, Outcome Binding, and total binding.

Half of the participants engaged in the task in the Direct condition followed by the Indirect condition; the other half engaged in the task in the Indirect condition followed by the Direct condition.

For each of the direct and indirect condition blocks, the Action-Outcome factor and the Baseline-Operant factor were counterbalanced using the Latin square method.

E. SOA

For SOA measurement, participants answered the following two questions after 12 trials under the following conditions [13][14].

1) SOA for control: In the conditions in which the action operation is performed, specifically the Baseline × Action, Operant × Action, and Operant × Outcome conditions, participants were asked, "How much did you feel you had control over your pointing?" and responded on a seven-point scale ranging from not at all to very strongly.

2) SOA for causality: In the conditions in which sound feedback on the action was given, specifically the Operant × Action and Operant × Outcome conditions, participants responded on a seven-point scale to the question, "How much did you feel that your button press caused the sound to beep?"

F. Action, Outcome, and Total Binding

Three types of binding are defined, as shown in Figure 3.

Generally, the perception of the timing of an action is delayed in a situation where there is feedback (Operant × Action condition), as opposed to a situation where there is no feedback (Baseline × Action condition). This delay is called action binding.

Similarly, the perception of the timing of the sound is brought forward when an action is performed (Operant × Outcome condition), as opposed to a situation in which there is no action (Baseline × Outcome condition). The time carried forward is called outcome binding.

The total of the action binding and the outcome binding is the total binding.

III. RESULT

A. Performance

Figure 4 (a) shows the completion time, that is, the duration from the start of the task through the time that the final number 3 was tapped. Figure 4 (b) shows the distance traveled by the cursor on the tablet.

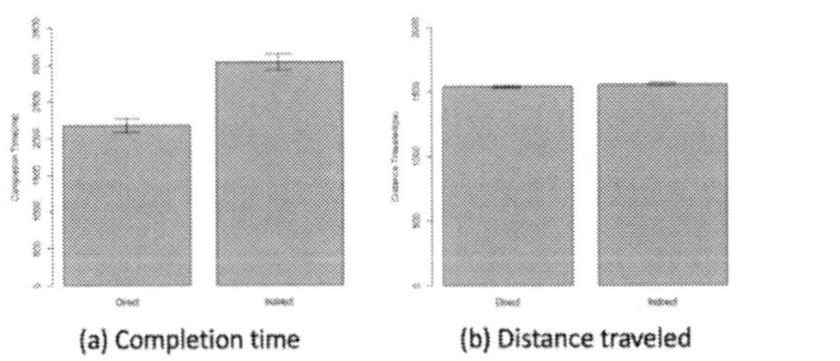

Figure 4. Completion time and distance traveled.

The completion time in the Direct condition was significantly shorter than that in the Indirect condition ($t(15) = 12.78$, $p = .001$), and the distance traveled in the Direct condition was marginally significantly smaller than that in the Indirect condition ($t(15) = 1.82$, $p = .09$).

B. SOA

Figure 5 (a) shows SOA for control and Figure 5 (b) shows SOA for causality.

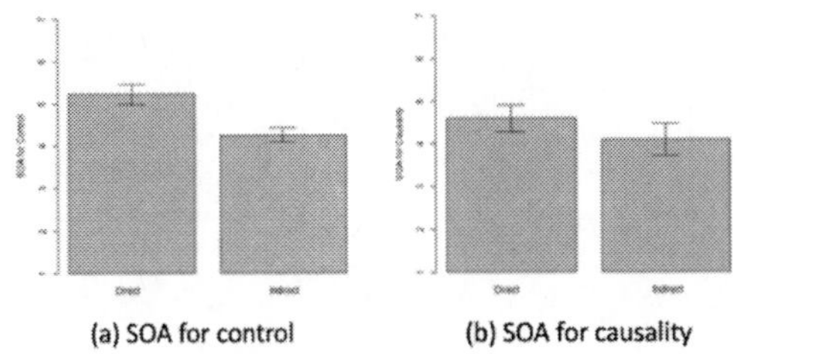

Figure 5. SOA for control and SOA for causality.

The SOA for control in the Direct condition was significantly larger than in the Indirect condition ($t(15) = 2.63$, $p = .002$), but the SOA for causality shows no significant difference between the Direct and Indirect conditions ($t(15) = 0.28$, n.s.)

C. Binding

Figure 6 (a) shows the action, outcome, and total bindings in the Direct condition, while Figure 6 (b) shows the three bindings in the Indirect condition.

The expected effects in all three bindings were confirmed under both the Direct and Indirect conditions.

D. Comparison of Binding

Figure 7 (a) shows a comparison of the action bindings in the Direct and Indirect conditions, Figure 7 (b) shows a comparison of the outcome bindings, and Figure 7 (c) shows a comparison of the total bindings.

No significant differences were detected between the Direct and Indirect conditions in action binding, outcome binding, or total binding ($t(15) = 0.21$, n.s. for action binding, $t(15) = 0.20$, n.s. for outcome binding, $t(15) = 0.15$, n.s. for total binding).

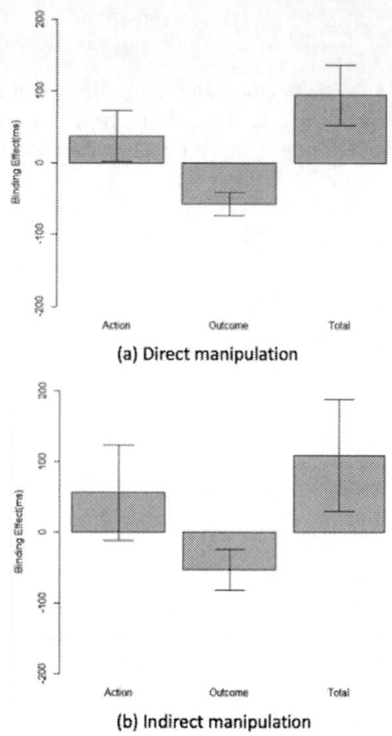

Figure 6. Action, outcome, and total bindings

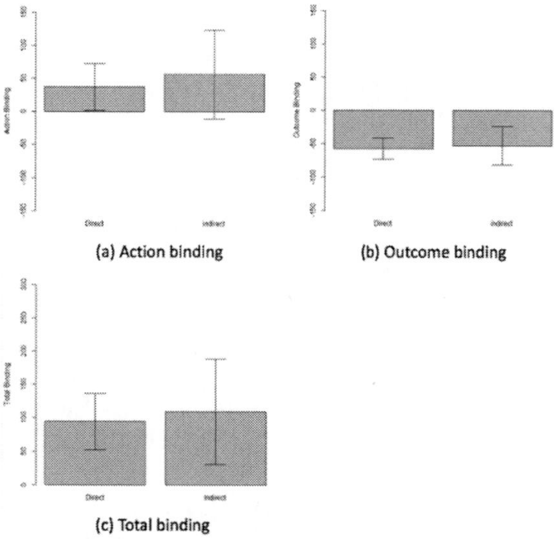

Figure 7. Comparison of action, outcome, and total bindings in the Direct and Indirect conditions.

IV. Discussion and Conclusions

Intentional binding was measured by using a serial manipulation task. Two situations were set up: one in which the tablet was manipulated directly and the other in which the tablet was manipulated indirectly, using another input device. Direct manipulation showed a higher manipulation performance than indirect manipulation.

The experiment results show that expected binding was detected for both direct and indirect manipulation situations. However, there was no difference in the magnitude of the bindings between the two situations.

Regarding the subjective evaluation of SOA, direct manipulation exceeded indirect manipulation in terms of the evaluation of the sense of control. However, there was no difference in causal perception between the two situations. This suggests that the magnitude of bindings may be related to causal perception rather than the perception of the sense of control. We investigate this point using an experimental paradigm that allows us to control for causal perception, such as delaying the feedback of the action [15][16].

References

[1] G. Prabhakar and P. Biswas, "A brief survey on interactive automotive ui," Transportation Engineering, vol. 6, 2021, p. 100089.

[2] S. Seinfeld, T. Feuchtner, A. Maselli, and J. Müller, "User representations in human-computer interaction," Human–Computer Interaction, vol. 36, no. 5-6, 2021, pp. 400–438.

[3] J. W. Moore, "What is the sense of agency and why does it matter?" Frontiers in psychology, vol. 7, 2016, p. 1272.

[4] H. Limerick, D. Coyle, and J. W. Moore, "The experience of agency in human-computer interactions: a review," Frontiers in human neuroscience, vol. 8, 2014, p. 643.

[5] K. Kilteni, R. Groten, and M. Slater, "The sense of embodiment in virtual reality," Presence: Teleoperators and Virtual Environments, vol. 21, no. 4, 2012, pp. 373–387.

[6] J. A. Dewey and G. Knoblich, "Do implicit and explicit measures of the sense of agency measure the same thing?" PloS one, vol. 9, no. 10, 2014, p. e110118.

[7] N. Khalighinejad and P. Haggard, "Extending experiences of voluntary action by association," Proceedings of the National Academy of Sciences, vol. 113, no. 31, 2016, pp. 8867–8872.

[8] P. Haggard, S. Clark, and J. Kalogeras, "Voluntary action and conscious awareness," Nature neuroscience, vol. 5, no. 4, 2002, pp. 382–385.

[9] J. W. Moore and S. S. Obhi, "Intentional binding and the sense of agency: a review," Consciousness and cognition, vol. 21, no. 1, 2012, pp. 546–561.

[10] P. I. Cornelio Martinez, S. De Pirro, C. T. Vi, and S. Subramanian, "Agency in mid-air interfaces," in Proceedings of the 2017 CHI Conference on Human Factors in Computing Systems, 2017, pp. 2426–2439.

[11] D. Coyle, J. Moore, P. O. Kristensson, P. Fletcher, and A. Blackwell, "I did that! measuring users' experience of agency in their own actions," in Proceedings of the SIGCHI conference on human factors in computing systems, 2012, pp. 2025–2034.

[12] J. Bergstrom-Lehtovirta, D. Coyle, J. Knibbe, and K. Hornbæk, "I really did that: sense of agency with touchpad, keyboard, and on-skin interaction," in Proceedings of the 2018 CHI Conference on Human Factors in Computing Systems, 2018, pp. 1–8.

[13] A. Kalckert and H. H. Ehrsson, "Moving a rubber hand that feels like your own: a dissociation of ownership and agency," Frontiers in human neuroscience, vol. 6, 2012, p. 40.

[14] N. Braun, J. D. Thorne, H. Hildebrandt, and S. Debener, "Interplay of agency and ownership: the intentional binding and rubber hand illusion paradigm combined," PloS one, vol. 9, no. 11, 2014, p. e111967.

[15] A. Sato and A. Yasuda, "Illusion of sense of self-agency: discrepancy between the predicted and actual sensory consequences of actions modulates the sense of self-agency, but not the sense of self-ownership," Cognition, vol. 94, no. 3, 2005, pp. 241–255.

[16] M. A. F. Ismail and S. Shimada, "'robot' hand illusion under delayed visual feedback: Relationship between the senses of ownership and agency," PloS one, vol. 11, no. 7, 2016, p. e0159619.

A Neural Network-Based Estimation of Tire Self-Aligning Torque

Younesse El Mrhasli
Bruno Monsuez
Department of Computer and System Engineering
Ensta Paris, Institut Polytechnique de Paris
Palaiseau, France
E-mail: {younesse.el-mrhasli, bruno.monsuez}@ensta-paris.fr

Xavier Mouton
Chassis Systems Department
Groupe Renault
Guyancourt, France
E-mail: xavier.mouton@renault.com

***Abstract*—Virtual sensing has attracted the interest of car makers and automotive service providers, owing to its cost-effective advantages, capacity to extract valuable insights from car data and its significance in enhancing the reliability of Advanced Driving Assistance Systems (ADAS). For instance, accurate virtual sensing of tire forces and torques can help adapt and improve the control strategies embedded in the vehicle's active safety systems. This paper deals with tire Self-Aligning Torque (SAT) estimation, an inherent parameter for identifying the limits of the vehicle at an early stage to prevent skidding. We present a data-driven approach to estimate the right and left front SATs, using a Neural Network (NN) model. The estimator takes directly existing in-vehicle signals and does not rely on expensive and unpractical sensors, which makes it cost-efficient and fast. Simulation results based on a high-fidelity vehicle model show a good performance of the chosen NN to estimate the SATs while considering the combined slip and road friction change.**

***Index Terms*—Tire Self-Aligning Torque, Estimation, Neural Network, Simulation.**

I. Introduction

To improve vehicle handling and ensure passenger safety, current research trends of Advanced Driving Assistance Systems (ADAS) and Automated Driving (AD) are focusing on monitoring the vehicle states, computing the road friction conditions, and adapting the control outputs according to the identified situation. Since the physical interaction between the car and the road occurs through the tire, estimating the forces and the moments applied at the contact surface of the tire in real-time is essential for developing advanced, performance-oriented, and safe driving assistance or automated driving systems [1]–[3]. For active safety systems, real-time identification of the maximum grip μ on the road is a critical task. Estimating the tire self-aligning torque (SAT), *i.e* the torque that a tire creates as it rolls along its vertical axis, allows to detect when the vehicle reaches its maximal lateral and longitudinal force capacity before the skid: it peaks at a lower slip angle than that corresponding to the maximum of the lateral forces (FY) (Figure 1). However, only a few contributions are harnessing this physical characteristic of the SAT. Current SAT estimation can be classified into two categories: the estimation based on an analytical model and the model-less estimation. Estimations based on analytical models use a physical or empirical tire model to infer the SAT. On the other hand, the model-less approach does not need an explicit tire model to build the virtual sensor.

The present study belongs to the second category and proposes using a Neural Network (NN) model to directly and cost-effectively estimate SAT with the aid of already existing sensors, along with left and right suspension deflection sensors. The latter has gained popularity in various applications such as vertical parameter estimation [4], [5], and skyhook control due to its cost-effective nature.

The structure of this paper is as follows: in Section II, we review existing methods for estimating the SAT and evaluate their performance. In Section III, we introduce the NN-based approach that we use for SAT estimation. Section IV presents the results of our simulations and provides an interpretation of the NN model. In Section V, we discuss the potential applications of SAT estimation. Finally, in Section VI, we outline future work to enhance the robustness of our observer and validate it on real data, before concluding with a summary of our findings.

II. Related work

Concerning the analytical model approach, Lenzo et al. [6] successfully estimate the SAT from a Brush tire model. First, their method uses the TRICK tool (Tyre/Road Interaction

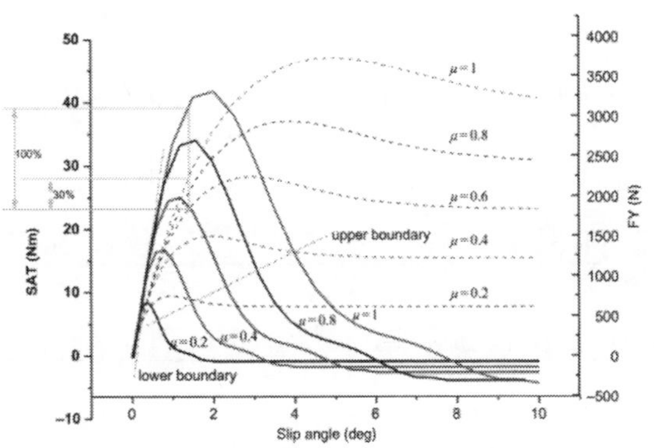

Fig. 1. SAT and Lateral force (FY) vs slip angle at different friction coefficients μ; (SAT: solid line , FY: dashed line).

Characterization & Knowledge) [7] to estimate the lateral forces. Then, the parameters of the Brush model are optimized to fit the estimation and are used to compute the SAT. The effectiveness of this approach depends on the accuracy and convergence speed of the TRICK tool.

Model-less estimation is mainly a data-driven method based on dedicated sensors such as force transducers, tie rod forces sensors [8] or sensors mounted on the kingpins [9]. Pasterkamp and Pacjeka [9] present a 3-layer NN fed by the steering wheel angle, the suspension inclination angle, the forces on the kingpins and the force in the steering link to estimate the forces and the SAT. Despite accurate results, the training and validation test cases were not extensive. In addition to that, sensors used to map the non-linearities are not commonly mounted on commercial vehicles. Luque et al. [8] employed a 2-layer NN (NN) to estimate the front right and left SATs in their study. The input to the NN consisted of tire longitudinal and lateral forces inferred from a Random Walk Extended Kalman Filter (RW-EKF), along with front axle vertical forces, steering wheel angle, and steering tie-rod forces measured by extensiometric sensors. However, one major drawback of this approach is that the error in the estimated forces from the RW-EKF, due to non-Gaussian noise, can be propagated to the outputs of the NN, resulting in decreased accuracy of SAT estimation.

III. Self-Aligning Torque Estimation

A. *Data Acquisition and Context of Study*

In our case, we use a high-fidelity vehicle model from AMESIM software, equipped with Electric Power Assisted Steering (EPAS) system and two suspension deflection sensors mounted on the front right and left.

To extract sufficient and reliable data and to map our entries to different regions of the SAT, the simulation was done on different open loop handling maneuvers, as depicted in Table I.

TABLE I
OPEN LOOP MANOEUVERS DONE IN SIMULATION.

ISO Maneuver	Longitudinal velocity range (Km/h)
Free steer	20 – 80
Steering pulse	20 – 80
Double lane change	40 – 120
Circular maneuver	40 – 80
One transient	40 – 80
Random swept sine steer	40 – 80
Braking in a turn	40 – 120
Fishhook	40 – 80
Sine with dwell	40 – 120
Steady brake/acceleration command	40 – 120

It is worth mentioning that the maneuvers were simulated with high repeatability on dry asphalt and other grip surfaces ranging between 0.7 (wet) to 0.2 (ice). Moreover, this evaluation did not consider active safety systems such as the Anti-lock Braking System (ABS) or the Electronic Stability Program (ESP).

We consider some measurable inputs related to the steering system [10] and other vehicle dynamic-related signals: The first part consists of choosing the steering wheel angle, the steering torque, and the assist torque according to the equation (1) of a second order steering system model. These measurements are available if the car has EPAS.

$$J_{eff}\ddot{\delta} + b_{eff}\dot{\delta} = \tau_{SAT} + \tau_{SW} + \tau_{assist} - \tau_f \qquad (1)$$

where J_{eff} is the effective moment of inertia, b_{eff} is the effective damping of the steering system at the road wheels, and δ is the steering wheel angle. τ_{SAT}, τ_{SW}, τ_{assist}, and τ_f represent the Self-Aligning Torque (SAT), the steering wheel torque, the assist torque, and the frictional torque at the road wheel, respectively.

The SAT observed from the previous equation is different from the real one. The main reason is the complexity of the tire behavior [9] due to the variation of the load, the couplings between longitudinal and lateral slips, and the non-linearities due to suspensions. To take this into account, additional measurable signals are considered such as the longitudinal and lateral accelerations, the longitudinal velocity, the yaw rate, the wheels speed, the wheel torque, and the compression/decompression of front right and front left suspensions. In total, 12 inputs are used to train the neural network to estimate the front right and the front left SATs. Specifications of the input and output data are listed in Table II.

TABLE II
INPUT & OUTPUT DATA.

Inputs		
Longitudinal acceleration Lateral acceleration	A_x (ms^{-2}) A_y (ms^{-2})	
Longitudinal velocity	V_x (ms^{-1})	
Yaw rate Steering angle Steering torque	$\dot{\psi}_z$ (rads^{-1}) $\alpha_{steering}$ (rad) $\tau_{steering}$ (Nm)	 from EPAS
Assist torque	τ_{assist} (Nm)	from EPAS
Motor torque	τ_{motor} (Nm)	
Compression/Decompression of the suspensions	$coslad_{left}$ (m) $coslad_{right}$ (m)	from front left sensor from front right sensor
Wheel speed	ω_{left} (rads^{-1}) ω_{right} (rads^{-1})	from front left sensor from front right sensor
Outputs		
Self-Aligning torque	τ^l_{SAT} (Nm) τ^r_{SAT} (Nm)	front left front right

From the previous remarks and due to the variation of the pressure distribution in the tire, the use of a physical tire model such as the Brush model is disregarded. Thus, we choose to label our data using the Pacjeka tire model or the Magic Formula [11]. This semi-empirical model fits best the measured data and takes into account the couplings between longitudinal and lateral slips. The details of SAT formula from Pacjeka 97 tire model can be found in [11].

In the first step, the correlation between all variables is performed to assess the dependency between the inputs and the outputs, as shown in Figure 2. The assist torque has the highest correlation value since it is linearly related to the SAT, as described in the steering system model equation (1). In addition, we also notice a medium dependency on suspension deflection sensors, highlighting the relation between the load variation and the SAT.

The data were sampled at 20 Hz, giving us an input matrix of (45000x12) and an output matrix of (45000x2).

The next part of this section will focus on the choice of the network model, the tuning of its parameters, and the definition of the performance metrics for evaluation.

B. Proposed Model

A static feedforward neural network or Multi-Layer Perceptron (MLP) is considered in this study. The goal is to use the MLP as a non-linear function approximator to map the entries to the SAT. In general, an MLP is composed of one input layer, one or more layers called hidden layers, and one output layer. The inputs of each layer are combined in a weighted sum and subjected to an activation function. Then, the result of this combination is propagated to the next layers. A backpropagation learning mechanism allows finally to adjust weights with the goal of minimizing the cost function.

The design of the NN model and the tuning of the hyper-parameters was done in an iterative manner using the Grid Search library in Python. This tool enables us to find the optimal hyperparameters by evaluating different combinations of values based on a defined performance metric. To assess the score of our predictor, we choose to use the R-squared metric defined as:

$$R^2 = 1 - \frac{\sum_{i=1}^{n}(y_i - \hat{y}_i)}{\sum_{i=1}^{n}(y_i - \bar{y})} \quad (2)$$

where n is the total number of measurements, y_i is the true measured value, $\hat{y}_i$ is the predicted value and $\bar{y}$ is the average of all measures. The best possible score for R^2 is 1.

The optimal model has 2 hidden layers with 32 and 12 neurons, respectively. We use the hyperbolic tangent activation function for non-linear mapping and the Adam optimization [12] for training. Table III summarizes the set of the chosen hyperparameters and the estimation structure is presented in Figure 3.

TABLE III
OPTIMAL PARAMETERS FOR THE MLP.

Parameter	Optimal
Hidden dimensions	[32,12]
Learning rate	Adaptive
Optimizer	ADAM
Activation function	Hyperbolic tangent
Data pre-processing	Robust Scaler

IV. SAT ESTIMATION RESULTS

A. Simulation results

To test the performance of our model, the recorded data were randomly split into 70% for training and 30% for testing. The optimal NN model yields an R-squared score of 0.986 for the first and 0.982 for the second. The Mean Absolute Error (MAE), which is less sensitive to the outliers caused by software compilation errors, is found to be 2.4 (Nm) in training and 2.59 (Nm) in the test phase.

To appraise the extrapolation ability of our NN model, we run the same vehicle model on the Magny-Cours race track. This sort of track is available on Simcenter AMESIM and is generally used to simulate severe maneuvers. The reference trajectory, the longitudinal velocity, and the steering wheel angle of the simulation are depicted in Figure 4.

The car does two rounds, the first one on a dry surface (μ=1) and the second on wet asphalt (μ=0.7). The results of estimation on dry asphalt presented in Figure 5 show that our NN model predicts accurately the front wheels SATs with an MAE of 3.1 (Nm). The blue line represents the true value and the red one is the NN estimation. On the wet road though, the MAE increases to 8.1 (Nm) and the NN does poorly to extrapolate the peak of the SAT. The results of this second case are plotted in Figure 6. Table IV summarizes all simulations' values of MAE and R-squared.

TABLE IV
SUMMARY OF SIMULATION RESULTS.

Simulation test	R^2 score	MAE (Nm)
Training phase	0.986	2.4
Test phase	0.982	2.59
Magny-Cours dry asphalt	0.971	3.1
Magny-Cours wet asphalt	0.931	8.1

One last thing to highlight is the good accuracy of our NN model to estimate the total aligning moment of the front axle *i.e.* the sum of the front right and front left SATs, as shown in Figure 7.

This observation proves that our NN model would be accurate to target mainly the front axle maximum grip estimation. However, it will do less to predict a μ-split case for example.

B. Interpretation of the model

To boost our model transparency, we will provide its interpretation based on SHapley Additive exPlanations (SHAP) [13]. SHAP was introduced as a unified framework for interpreting predictions. It is a game theoretic approach that assigns each feature an importance value. Two types of explanations are accessible via SHAP: A global one where the SHAP values show how much each predictor contributes to the target variables. And, a local one dedicated to a specific observation.

In this paper, we will provide only global interpretability based on the test set data. Figure 8 is a bar plot that lists the most influencing features in descending order and the average impact on the SAT magnitude is shown on the x-axis. On the

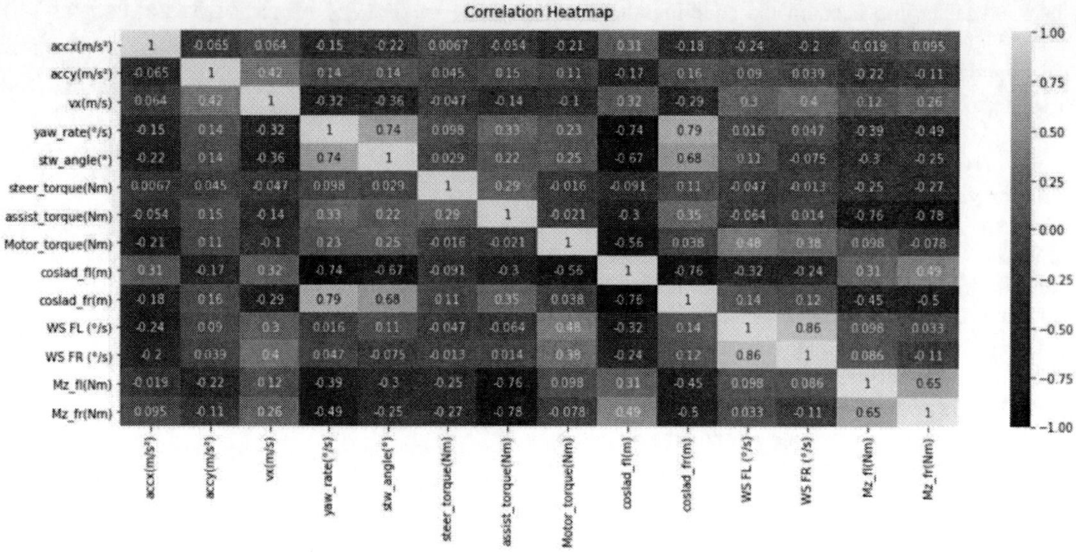

Fig. 2. Correlation matrix of input and output data.

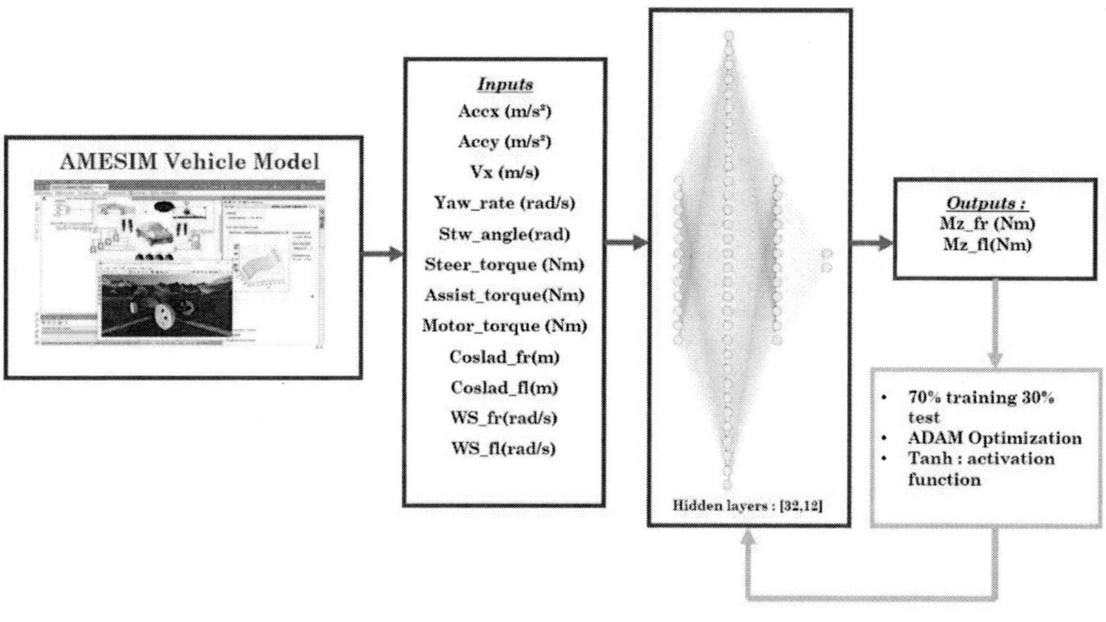

Fig. 3. SAT estimation structure, adapted from [8].

other hand, the dependence plot depicted in Figure 9 explains the marginal effect between the top 3 features and the front left SAT. From the latter, we observe a negative linear relationship between the assist torque and the front left SAT. While for lateral acceleration and the steering wheel angle, the effect on the SAT is non-linear.

V. Discussions

A. *SAT dependency on inflation pressure*

Tire inflation pressure has an influence on the quasi-static generated forces and moments, most importantly, the SAT. From a physical perspective, the SAT is generated because of the distance between the contact patch center and the point of lateral force application, this distance is called the pneumatic trail and it is linearly dependent on the contact patch.

An investigation on the effect of pressure change on SAT was carried out using an extended version of the Pacjeka tire model in AMESIM environment. This model called SWIFT-Tyre has been developed at Delft University of Technology and TNO Automotive [14] and includes the most recent developments such as inflation pressure effects. We observe from Figure 10 that the amplitude of SAT decreases when

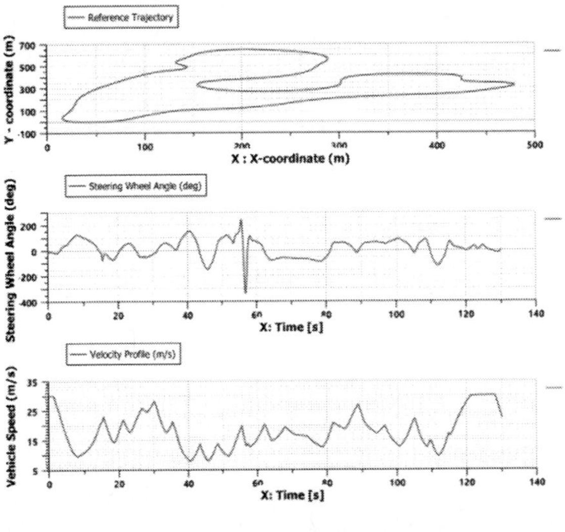

Fig. 4. Reference trajectory (top), steering wheel angle (middle), and velocity profile (bottom) for Magny Cours Track.

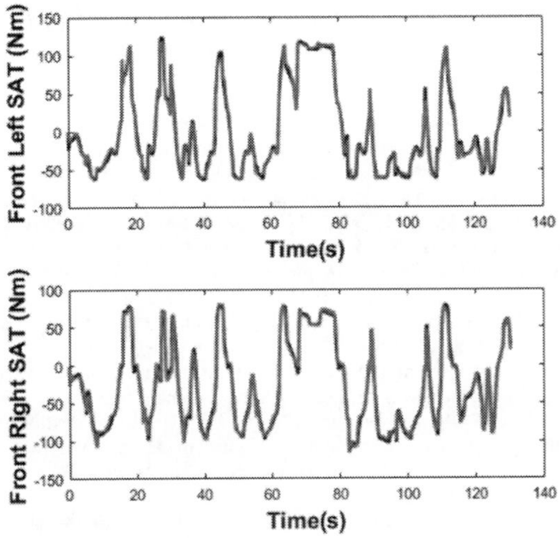

Fig. 5. Estimation results of SAT on dry asphalt; Blue (True) and Red (Estimated with NN).

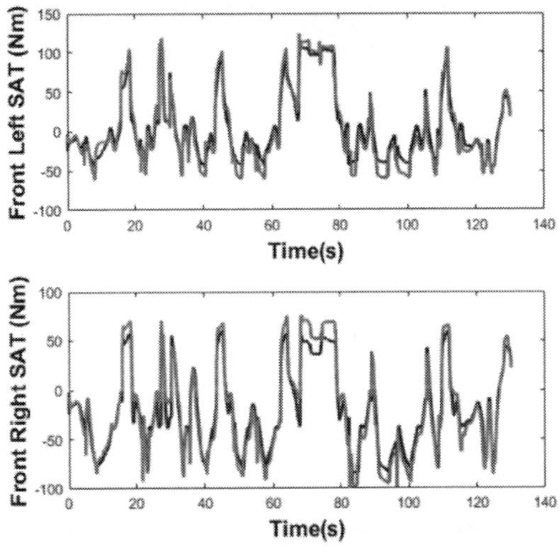

Fig. 6. Estimation results of SAT on wet asphalt; Blue (True) and Red (Estimated with NN).

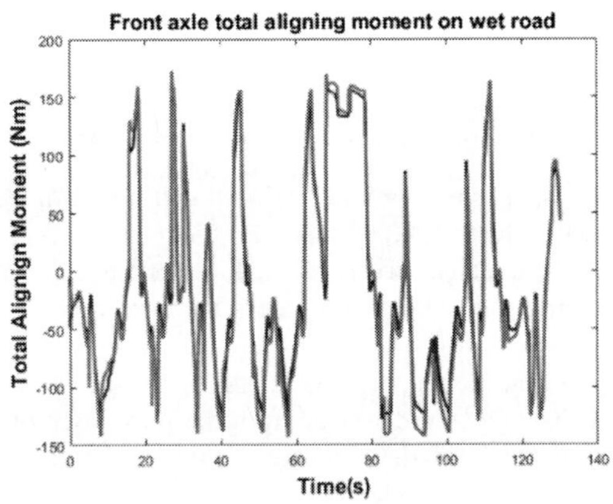

Fig. 7. Estimation of front axle total aligning moment on wet asphalt; Blue (True) and Red (Estimated with NN).

inflation pressure increases. This is logical because higher pressure reduces the contact length, thus the pneumatic trail decreases and eventually also the SAT. This leads us to consider in a future study the tire's inflation pressure acquired from Tyre Pressure Monitoring Systems (TPMS) as an input of our NN model to enhance the performance and robustness of our estimator. Or in a simpler way, consider a corrective term that will compensate for the effect of the pressure.

B. Applications of SAT estimation

What motivates most the SAT estimation is the early detection of tire friction coefficient. Unlike other traditional approaches that reach a good estimate near the critical region of the tire, the SAT is viable for limits detection at low excitation levels. Owing to this, the knowledge of friction conditions is prior to the intervention of advanced active safety systems.

The knowledge of SAT can also improve the lateral control [15] and particularly the Steering Wheel Angle (SWA) control in EPAS systems [16]. While it is considered a disturbance to be overcome in most controllers, a precise estimation can prevent generating inefficient control gains and cancel its effects in some situations. Moreover, it can be useful to return to the center position of the SWA after a change in direction.

To wrap up, real-time estimation of SAT is inherent to guarantee safety by providing the available grip at an early stage and also enhancing the performance of some lateral controllers.

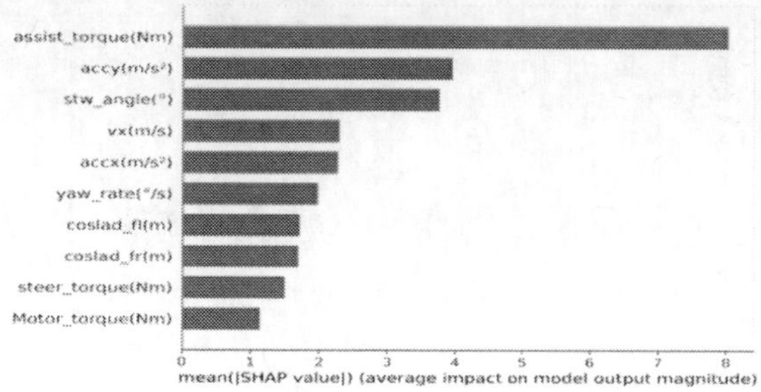

Fig. 8. Feature Importance based on SHAP.

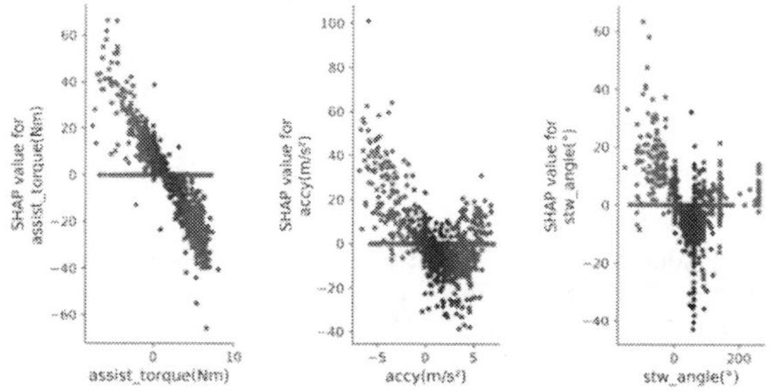

Fig. 9. Dependency plot of the top 3 features.

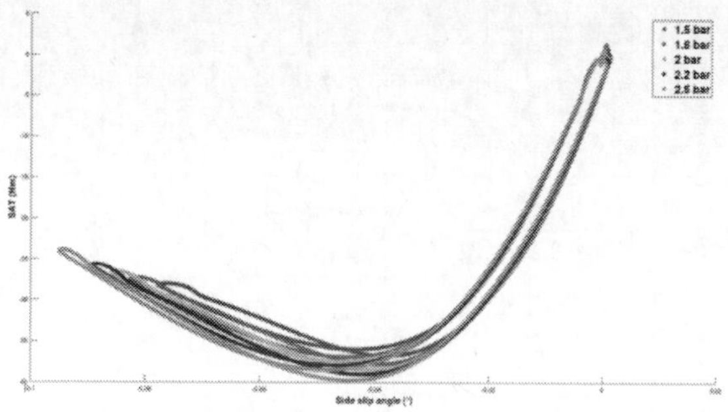

Fig. 10. Inflation pressure effect on SAT for a triangle-shape steer command.

VI. Conclusion and Future Work

In this paper, we presented and outlined a real-time data-driven approach for SAT estimation. This quantity is inherent for friction coefficient prediction at low excitation levels and for enhancing some lateral controllers' performance e.g (SWA control).

The proposed neural network model and the methodology followed distinguish themselves from the previously reported methods in terms of the following features: 1) The NN is fed directly by in-vehicle sensor signals and does not rely on estimated inputs nor uncommon expensive sensors. 2) It is trained and tested on a wide range of maneuvers with different road surfaces to improve its extrapolation ability. 3) Labeling the data uses a semi-empirical tire model (Pacjeka tire) that considers combined lateral and longitudinal dynamics and can fit the measured SAT on a real test drive. 4) A global interpretation based on SHAP values is provided. It gives us the most important features and the nature of their relationship with the estimated SAT. We investigated also the effect of inflation pressure on SAT by using an extended version of the same tire model, and we deduced that for more robustness and precision, the pressure acquired from TPMS can be considered as an additional input in our model.

The graphs and regression metrics show a good performance of our NN model to estimate the front right and front left SATs, especially for tests on dry asphalt. As the error increases for the wet road test, enriching the dataset with repeatable maneuvers on other grip surfaces may resolve this problem.

Future work will be oriented towards generating larger data sets in different friction coefficients and considering the inflation pressure as an input, in order to refine the generalization of our estimator. Besides, this estimator will be used for friction estimation in a subsequent paper. Finally, a real test drive is planned with GROUPE RENAULT for validation and evaluation.

References

[1] M. Viehweger *et al.*, "Vehicle state and tyre force estimation: demonstrations and guidelines," *Vehicle System Dynamics*, vol. 59, no. 5, pp. 675–702, 2021.

[2] X. Jin, G. Yin, and N. Chen, "Advanced estimation techniques for vehicle system dynamic state: A survey," *Sensors*, vol. 19, no. 19, 2019.

[3] M. Acosta, S. Kanarachos, and M. Blundell, "Road friction virtual sensing: A review of estimation techniques with emphasis on low excitation approaches," *Applied Sciences*, vol. 7, no. 12, 2017.

[4] B. Wang, H. Wang, L. Wu, L. Cai, D. Pi, and E. Wang, "Truck mass estimation method based on the on-board sensor," *Proceedings of the Institution of Mechanical Engineers, Part D: Journal of Automobile Engineering*, vol. 234, no. 10-11, pp. 2429–2443, 2020.

[5] B. L. Boada, M. J. L. Boada, and H. Zhang, "Sensor fusion based on a dual kalman filter for estimation of road irregularities and vehicle mass under static and dynamic conditions," *IEEE/ASME Transactions on Mechatronics*, vol. 24, no. 3, pp. 1075–1086, 2019.

[6] B. Lenzo, S. A. Zingone, and F. Timpone, "On the estimation of tyre self-aligning moment through a physical model and the trick tool," *International Journal of Mechanics and Control*, vol. 21, no. 2, pp. 13–20, December 2020.

[7] F. Farroni, "T.r.i.c.k.-tire/road interaction characterization & knowledge - a tool for the evaluation of tire and vehicle performances in outdoor test sessions," *Mechanical Systems and Signal Processing*, vol. 72-73, pp. 808–831, 2016.

[8] P. Luque *et al.*, "Tyre–road grip coefficient assessment – part ii: online estimation using instrumented vehicle, extended kalman filter, and neural network," *Vehicle System Dynamics*, vol. 51, no. 12, pp. 1872–1893, 2013.

[9] W. R. Pasterkamp and H. B. Pacejka, "The tyre as a sensor to estimate friction," *Vehicle System Dynamics*, vol. 27, no. 5-6, pp. 409–422, 1997.

[10] Y.-H. J. Hsu, S. M. Laws, and J. C. Gerdes, "Estimation of tire slip angle and friction limits using steering torque," *IEEE Transactions on Control Systems Technology*, vol. 18, no. 4, pp. 896–907, July 2010.

[11] H. B. Pacejka and I. J. M. Besselink, "Magic formula tyre model with transient properties," *Vehicle System Dynamics*, vol. 27, no. sup001, pp. 234–249, 1997.

[12] D. Kingma and J. Ba, "Adam: A method for stochastic optimization," *International Conference on Learning Representations*, 2014.

[13] S. M. Lundberg and S. Lee, "A unified approach to interpreting model predictions," *CoRR*, vol. abs/1705.07874, 2017.

[14] A. J. Schmeitz, I. J. Besselink, J. De Hoogh, and H. Nijmeijer, "Extending the Magic Formula and SWIFT tyre models for inflation pressure changes," *VDI Berichte*, no. 1912, pp. 201–225, 2005.

[15] Z. Yu, Y. Hou, B. Leng, L. Xiong, and Y. Li, "Disturbance compensation and torque coordinated control of four in-wheel motor independent-drive electric vehicles," *IEEE Access*, vol. 8, pp. 119 758–119 767, 2020.

[16] W. Kim, C. M. Kang, Y.-S. Son, and C. C. Chung, "Nonlinear steering wheel angle control using self-aligning torque with torque and angle sensors for electrical power steering of lateral control system in autonomous vehicles," *Sensors*, vol. 18, no. 12, 2018.

Towards Cross-Cultural Intelligent Vehicles: A Review

Imane Taourarti
Research Department / Autonomous Systems and Robotics Lab,
Computer Science and System Engineering (U2IS)
Renault Group / ENSTA Paris,
Institut Polytechnique de Paris
Guyancourt, France
email: imane.taourarti@{renault.com, ensta-paris.fr}

Arunkumar Ramaswamy
Javier Ibanez-Guzman
Research Department
Renault Group
Guyancourt, France
email: arunkumar.ramaswamy@renault.com,
javier.guzman-ibanez@renault.com

Bruno Monsuez
Adriana Tapus
Autonomous Systems and Robotics Lab,
Computer Science and System Engineering (U2IS)
ENSTA Paris, Institut Polytechnique de Paris
Palaiseau, France
email:{bruno.monsuez, adriana.tapus}@ensta-paris.fr

***Abstract*—Advanced Driver Assistance Systems (ADAS) are becoming an integral part of modern road vehicles. Their deployment is demonstrating their contribution to safety and efficiency. However, as the interaction between ADAS and the driver increases, other issues are emerging that affect their performance. The driving task is influenced by a range of factors, including the driver's preferences and behavior that is conditioned by the operating environment comprising the road conditions, environmental conditions, and complex social interactions with other road users and pedestrians, etc. Driving differs also between and within cultures. In this paper, we review the current approaches in the literature that demonstrate an adaptation to the driver behavior but also the work on social interactions on the road. We then discuss issues that remain open and need to be confronted when designing a cross-cultural intelligent vehicle.**

***Index Terms*—intelligent vehicles, culture, context-based system, safety, personalized ADAS, social robotics.**

I. Introduction

Vehicles and driving are intimately connected to our individual and collective sense of self - who we are, what we believe in, what are our values, and what we aspire to achieve, as well as how we interact with others [1]. Currently, there is a rapid deployment of Advanced Driver Assistance Systems (ADAS) in the new generation of road vehicles as a means to enhance safety, riding comfort, and energy consumption. Their deployment is contributing to improvements in these areas, with modern legislation and vehicle qualification evaluations such as the EuroNcap [2]. However, the interaction between ADAS systems and drivers is becoming very symbiotic, which raises several issues.

Intelligent vehicles are mainly developed based on data collected, developments and field trials, and research conducted in North America, certain countries in Asia and Europe, where driving conditions, safety, etc. are very different from what occurs elsewhere [3]. It is to be noted that in low- and middle-income countries a growing phenomenon is occurring, road accidents are reaching almost epidemic proportions, and road safety has became a major concern. The World Health Organization [4] reported that with an average rate of 27.5 deaths per 100,000 population, the risk of a road traffic death is more than three times higher in low-income countries than in high-income countries, where the average rate is 8.3 deaths per 100,000 population, see Fig. 2. Furthermore, these countries have also witnessed a major increase in the number of road vehicles. In these countries, the road infrastructure, traffic conditions, driver training, and respect to the traffic code are substantially different [5].

The conditions and road networks where ADAS functions are deployed differ very much. Recently, research has increasingly focused on reducing bias in the development of intelligent vehicles by addressing the intricacies raised by cultural and social differences [3], [6]. In a developing country such as India, Fig. 1, in order to respond to common challenges on the road, traffic conditions, local regulations, and unwitting rules are rapidly emerging. For example, in heavy traffic, respect to the rule that should keep all cars within the boundaries of lane markings disappears, that is more cars than the number of lanes will fit across standard roads. Unlike countries within the European Union, there will be more non-verbal cues and verbal communication to create awareness and for drivers to find a consensus related to safety and efficiency. Another example would be crossing outside crosswalks that is a common behavior of vulnerable road users; this is contrary

Fig. 1. Sample of traffic conditions in India [8]–[10].

to what occurs in most Nordic countries [7]–[10].

The differences in road networks and operating conditions are more notable when deploying the Society of Automotive Engineers (SAE) Level 4 vehicles (e.g., robot-taxis); that is, the machine should understand its situation before making any decision; however, situations will vary from country to country, from rural land to dense urban areas, even within the same country. Most field trials of robot-taxis have been so far confined to limited areas, and scaling up has proved more difficult than expected. Therefore, the manner of how all road users will behave is a major constraint to full-scale deployment.

Intelligent vehicles comprising ADAS functions or different levels of automation will not achieve their promise if drivers and the environment rounding do not accept and use them in a sustainable manner [11].

Designing a cross-cultural intelligent vehicle is one of the challenging problems faced by researchers in the automotive sector but not yet seriously addressed. Through this paper, we tackle the following questions:

- What do we mean by a culture with respect to driving? What are the main cultural differences in driving behaviors?
- What do we mean by a cross-cultural adaptive intelligent vehicle? What are the cues to pay attention to on the road?

The remainder of this paper is organized as follows. In the following section, Section II, we define culture, introduce its dimensions in the context of driving, and we emphasize the bias in the development of intelligent vehicles to date. Following that, in Section III, we discuss personalization in Advanced Driver Assistance Systems and drivers models in the literature. In Section IV, we will review the state of the art in social intelligent vehicles and discuss where attention is turned to on the road. Section V discusses what a context-based intelligent vehicle would consider and highlights challenges and open issues in the design of such a system. Conclusion and future work are drawn in the last Section VI.

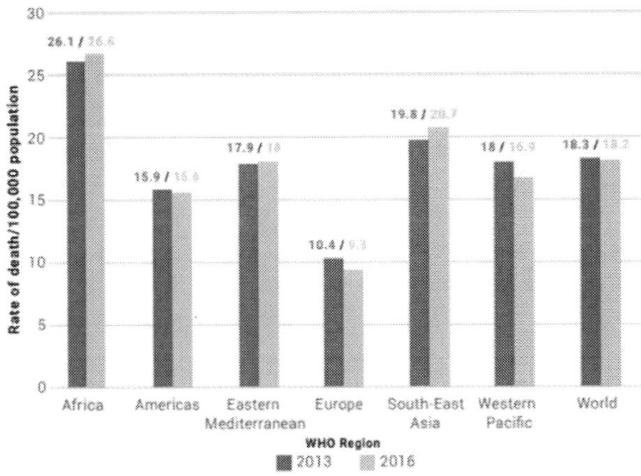

Fig. 2. Rates of road traffic death per 100,000 population between 2013 and 2016 by the World Health Organization regions [4].

II. CULTURE'S IMPACT ON DRIVING TASK AND ROAD TRAFFIC

The driving task is not only what is measured objectively on a road, nor only the professional conceptualization of the traffic system. It is also the "world view" that lay people have of the traffic system [12].

It is in the human being as a mirror of one's personality, one's expectations and risk assessment, also one's culture. In this context, How could intelligent vehicles gain their place in society and be accepted by the driver? The intelligent vehicle is involved in more than just trajectory optimization, obstacle avoidance, and driver cognition testing. To become a part of the surroundings and to incorporate the driving culture, it needs also to be merged with the driver and represent him/her.

Before diving into the role of culture in the driving task and why it is primordial to include the context in the design of intelligent vehicles, we first define the concept of culture.

In "What is Culture?" by Edgar Schein [13], culture is defined as a pattern of shared assumptions (knowledge and values) that have served a group well in the past, that is learned to new members and that can be adapted to external circumstances. Culture is essentially a social indoctrination rule that people learn as they try to fit into a particular group.

Culture can also be defined through its characteristics. It has collective representations - vocabularies, symbols, and codes [14], [15]. More often on some roads and less on others, drivers, to find a consensus, tend to communicate verbally their decisions and understand common non-verbal cues, while in Europe for example, we notice a certain individualism as rules are respected and traffic is structured. The work in [16] and [17] explicitly addresses the need for an external HMI for autonomous vehicles to communicate with other vehicles and pedestrians. Culture has social norms and values. All these elements structure the thinking and acting of the individuals within the same group to respond to the survival challenges of their environment by learning through reward and punishment, by conforming to social norms, laws, and regulations, by accepting persuasive messages, anticipating others' behavior, and compensating for the errors of other traffic participants.

As illustrated in Section III and concluded in [1], the ADAS systems can compensate for the perceptual constraints that affect driver performance when responding to roadway demands, but less when overriding a driver's attitudes, goals, and priorities. Driver behavior, then, may ultimately have the most influence on traffic safety. Research on driver behavior has focused almost entirely on individual differences (e.g., cautious, aggressive), distraction, or cognitive level as contributors to unsafe driving behavior. However, in [18], the authors investigate the relationship between the three factors of the Driver Behavior Questionnaire (DBQ) (errors, aggressive violations, ordinary violations) reflecting culture differences and the difference in the rate of fatalities in six countries (Finland, Great Britain, Greece, Iran, The Netherlands, and Turkey). Findings demonstrated that the addition of driving styles, especially aggressive violations and errors not only improved the models for predicting the number of traffic accidents but also mediated the relationship between culture/country and accidents. The results show that 84.6% of traffic accidents are caused by vehicle violations, which is the crucial factor within all traffic accidents [15]. In the road safety annual report in 2019 [19], it is emphasized that traffic-related mortality rates differ widely between countries, e.g., the risk of being killed in a road crash is six times higher in Argentina than in Norway. Although traffic car accidents are a major problem everywhere, significant differences between countries are encountered. The results emphasize the critical role that culture plays in driving safety. The authors in [20] discuss road accidents related to the interactions among drivers instead of single attitudes. A recent review [21] discusses the factors influencing driving behavior and the causes of road accidents.

Moreover, culture provides the subtext to driver behavior by shaping the beliefs, values, and ideas that people bring to the driver's seat. It highlights the influence of societal expectations and practices among drivers from the same culture. For example, "honking" clearly reflects aggression in Scandinavia, whereas in Southern Europe and Iran, drivers use their horn frequently to give various messages, such as thanking other drivers. Furthermore, in Turkey, the speed of traffic flow on many roads is much higher than the speed limit. Consequently, drivers do not see their speeding as a serious offense as the Western Europeans might do. Thus, it is important to consider that traffic culture or context determines the criteria but also both formal and informal rules for acceptable driving style, and thus, develop nation-specific items for reflecting informal rules that reflect the cultural behavior in each country/environment [18]. Understanding the context is what will give the intelligent vehicle fluidity, motion involved in social exchanges, the socio-acceptance and it is what is going to increase the driving safety by anticipating other drivers behavior and making up for errors made by other traffic participants.

Transferring the ADAS technology as designed by automotive manufacturers from one culture to another can be problematic. In [22] and [23], research was conducted to indicate different cultural areas that need to be focused on when developing ADAS for China. One of the major problems in China is the complex traffic environment with congestion, motorized three-wheeled vehicles, and poor lane markings. It has been reported that ADAS such as Forward or Lateral Collision Warnings or Adaptive Cruise Control can be just annoying in such crowded environments where people obey authority norms less than social ones. Consequently, an ADAS that is of great value to the drivers of one country may be of less value than to those in another if not adapted.

The intelligent vehicles are integrated into hybrid roads where drivers tend to possess a model of their environment, allowing them to predict the intentions of road and non-road users, and thus, their safe driving is predetermined based on meeting the expectations of others. The intelligent vehicle should possess this set of knowledge, skills, and competencies to recognize, understand, and adapt to social and cultural differences.

III. Personalized Driving Assistance systems: State of the art

Due to the greater market penetration, the field of advanced driver assistance systems has grown to include aid functions that are increasingly complex but designed for the average driver or all drivers [24]. To assure the best user experience throughout such a wide range of use conditions and usage patterns, personalization techniques have been created. Personalized ADAS are developed by learning driver models from the observation of driver behavior and then parameterize the vehicle controllers to meet the personal driving style. In this section, we review recent work on the driver models and the personalized ADAS.

A. *Driver modelling*

Since their pioneering theoretical study of Automobile-driving, human driving behavior, by Gibson and Crooks in 1938, scholars have contributed to driver behavior mimic and driver psychology modeling [25].

The models proposed went from simplistic mathematical models to represent the correlation between the state metrics of the host vehicle (acceleration, relative speed, distance headway, etc.) [26]–[28] to more sophisticated models reflecting the internal mechanisms of the decision making that drivers must hold in their minds. The authors in [29] model the driver behavior in the ACT-R cognitive architecture. In [30], J.A. Michon discusses the driver behavior model types; behavioral (Mechanistic, adaptive-control, etc.) and psychological (motivational, cognitive, etc.). The authors of [31] review two hundreds models on driver behavior modeling.

Each driver is individually influenced by the social environment consisting of other road users, general social norms, traffic-related rules of conduct, and their representations [32]. The models proposed in the literature either suggest that a group of similar characteristics, or stereotypes, exist about a set of users [33], [34] or they are tailored to meet personal driving styles. However, driving style is supposed to vary in the degree to which it is shaped by both intrinsic (e.g., age, sex, experience, cognitive biases, and emotions [35]) and extrinsic (e.g., social context) factors [18], which are rarely considered. In the same sense, driver behavior modeling as proposed in the literature lacks a connection between models of individual driver behavior and the (presumably) resultant population behavior as reflected in traffic characteristics, informal rules conducted, or the accidentology level in the environment, which greatly influence the driving skills, the other main component of human factors in driving [36], [37]. [38] models the driver behavior along with demonstrating how the contextual information affects its performance.

B. Personalized ADAS

In this part, we are interested in defining the personalization of the ADAS as studied in the literature, revealing the key human/environment features considered in this personalization and presenting the process behind including those features in the loop. We have considered functions representing the three types of driver efforts: strategic (route planning), tactical (Adaptive Cruise Control [46], [47] – Lane Change Assistance), and operational (Forward Collision Warning) in the aim of identifying which features are relevant to each type/function and coming across models that have an eye in and out of the vehicle. A recent survey [48] and a review of personalization in ADAS and autonomous vehicles [49] concentrate on methods that combine individual driver models and controllers for designing personalized ADAS.

Table I summarizes some of the papers reviewed. In the personalization of the ADAS, we can distinguish between group-based and individual-based approaches to personalization. In the former case, drivers are assigned to one of a small number of representative driving styles (e.g., aggressive, cautious, etc.). In the latter case, the ADAS strategy tries to best reproduce the driving style of an individual driver [50]. In the table, we make also the point on the driver's characteristics relevant for the function, the environmental dynamic information, etc. We also refer to the methods used for Driver/context behavior recognition and the models used for Adapting. Finally, the personalization as demonstrated today, lacks a continuous learning of the human preferences or proposes that on demand with a recalibration of the process of personalization, we distinguish between the two approaches in the table.

We have presented some ADAS functions referring to self-driving capabilities (Adaptive Cruise Control), maneuver assistance (Lane Change Assistance), and monitoring capabilities (Lane Keeping Assistance) to show how they share some common features and differ on others to approach the driver modeling.

The use of neural networks or fuzzy logic (with capabilities of approximation, generalization, and self-learning) is suitable for modeling driver behavior with nonlinear characteristics. The research showed that artificial intelligence could offer some potential advantages in driver behavior analysis and modeling. However, the current studies present some limitations listed below:

- The driving style and the driver behavior are studied mostly from the control viewpoint, e,g., mimicking the acceleration/deceleration profiles [41]. Although, this operation is the result of different traffic situations, intra-individual differences, etc.
- Current customized personalized systems are mainly implemented through manually adjusting warning trigger thresholds for example, which would be less feasible for overall drivers as a certain domain expertise is required to set personal thresholds accurately and it becomes a tedious task as the number of ADAS is continuously increasing.
- Personalization techniques exploit individual drivers' data to build personalized models. Such an approach could learn personal behaviors but requires impractical large-scale individual data collection or the data are mostly based on simulation and not close to reality.
- We did not come across papers studying different functions under the same framework and this is problematic as the ADAS functions are increasingly added.
- The personalization of the ADAS to meet the driver's preferences and to mimic his behavior is bottom-up and when it is top-down, it is not validated.
- Artificial intelligence has proven its potential in modeling driver behavior. However, it presents some disadvantages when coming to the model stability, the computational load required and the complexity of it.
- Two approaches exist for the trajectory modeling: stochastic (Hidden Markov Models, Neural Networks, Fuzzy logic, etc.) and kinematic [42]. the stochastic modeling has proven its capability to approach different driver behaviors, it is flexible and accurate but lacks the physical meaning contrary to the kinematic modeling.

Despite the aforementioned efforts, it is missing the inclusion of the driver skills which are related to the environment from which he gains experience and constructs this toolkit to respond to the road needs. The objective of a driver model is to represent the process by which a driver transforms some perceived information about the driving situation into an action on the vehicle's actuators (steering wheel, pedals). We believe that regenerating this behavior is not mimicking the brake and acceleration profiles but understanding the why of these maneuvers as the driving task is about the driver's behavior toward a certain situation.

IV. Social Autonomous vehicles

Research on social robotics and in particular social autonomous vehicles is demonstrating the importance that play the social and the cultural dimensions when it comes to situation understanding [51], decision-making, and motion planning [52]. [53], [54] reveal the gaps in the development of autonomous vehicles navigating in uncertain environments and the lack of sufficiently detailed understanding of how humans interact in such conditions and how that understanding

TABLE I
REVIEW OF PERSONALIZED ADVANCED DRIVER ASSISTANCE SYSTEMS.

Function	Approach	Driver/context in the loop	Model-used	Driver/Context behavior recognition	Methods for adapting	The learning rate	Reference
Adaptive cruise control	Group-based	average, maximum and minimum of relative speed, time headway and jerk	learning-based	Self-organizing feature map neural network with K-means then PNN classifier	MPC as an upper-level Controller, and feedforward and PID for lower-level controller	On demand	[39]
Adaptive cruise control	Individual-based	Demographics: Age, sex, income level, educational level - location of the vehicle: distance to the lead vehicle, vehicle speed, longitudinal acceleration, road density, road type, weather - driver's behavior	learning-based	Regression model – decision tree model	-	-	[40]
Adaptive cruise control	Individual-based	Motion states of the leading vehicle and the host vehicle (vehicle speed, acceleration, accelerator pedal/throttle depression, brake pressure, relative distance/speed to lead vehicle, and Global Positioning System information)	Model-based	Self-learning algorithm based on RLS to identify the model parameters	a linear driver behavior model with a lower PID controller	Continuous learning	[41]
Lane change assistance	Individual-based	Max/average absolute value of steering-wheel angle, average steering-wheel angular velocity, standard deviation of steering-wheel angle, max/average absolute value of lateral acceleration, maximum absolute value of slip angle, maximum absolute value of yaw angle, maximum value of yaw rate, average value of yaw rate, and standard deviation of yaw rate	Rule-based and learning-based	Fuzzy c-means algorithm for classification then back-propagation (BP) neural network optimized by a particle swarm optimization (PSO) algorithm	a sinusoidal lane-change model	Continuous learning	[42]
Forward Collision Warning	Individual-based	Time headway, Time to collision, longitudinal speed of ego vehicle	Model-based	recursive least squares method for warning threshold	Adaptive algorithm	Continuous learning	[43]
Forward Collision Warning	Individual-based	Gas pedal position, range with neighboring vehicles, turn signal, yaw rate, longitudinal acceleration, velocity…	learning-based	Neural network – Support Vector Machine	Adaptive algorithm	Continuous learning	[44]
Route planning	Group-based	The vehicle's absolute motion, the vehicle's relative motion to surrounding vehicles and/or objects, Distance, Time, acceleration profile…	Model-based	HMM models – classifiers – fuzzy-based classifier - …		Continuous learning	[45]

might be quantified in computer models. Considering the lane change maneuver, scholars are formulating the problem as a non-cooperative game [55] when considering the social behaviors and the intentions of the surrounding vehicles [56], [57] while in [58], the authors are imitating the stimulus-based selective attention mechanism of human vision systems to recognize the lane changing intention of the surrounding vehicles. Based on a higher level of cognition, human drivers have this capability to pay attention to relevant information on the road related to their actual maneuvers, the authors in [59] review the modeling of where and when the drivers look on the road. Additionally, human drivers consider the stochastic variability in their interaction with vehicles/pedestrians, [60] is addressing this problematic at uncontrolled crosswalks. Learning to drive has emerged as an efficient alternative to hand-crafted rules, especially when considering interactive behaviors [61]–[66]. To tackle the social, ethnographic, and legal dilemmas in the urban environment, [67] offers insights into a new automated driving strategy by introducing a general learning-based framework based on maximum entropy inverse reinforcement learning and the Gaussian process. In [68], the authors introduced the learning by watching others framework enabling the vehicle to learn new skills in a new situation or geographic location, which finds its inspiration in the driver capabilities to fit in new environments and cultures by watching demonstrations from other drivers. To fit into an environment and to gain this social invisibility [69], drivers are learning from countless experiences by possessing this device of permanent memory, inferring, and experiential updating in addition to their event-related mechanism. The authors of [70] are discussing how to implement a cognitive computing framework for autonomous driving with selective attention and event-driven mechanism. A direct measure of performance for autonomous vehicles is their level of similarity to human drivers, and emulating human driver behavior just adds more challenges to countless ones.

V. DISCUSSION

In order for an intelligent vehicle to gain in intelligence, performance, robustness, and acceptability, we posit that it should take a culturally and socially cognizant path. We define cultural awareness as the capacity to infer such favorable actions based on knowledge about the driver and others' intent and behavior, of understanding road and non-road users' interactions and complying with mutually-accepted rules but also compensating for the uncertainties and non-stationarities, thus, formulating their social insights.

A generalized framework of an intelligent vehicle having an eye in and out could be a step up. It should consider the following features, which we regroup in three sub-contexts:

1) ***The vehicle with automation***
2) ***Interaction with the operating environment:*** all the elements constituting the operating environment are linked, see Fig. 3. People from the same culture develop distinct patterns of emotions, norms (informal rules, local perception of the law, etc.), and behaviors to deal with the survival challenges of their common environment (infrastructure, environmental conditions, etc.) figuring out a structure in the unstructured environment. The system should be able to reason about these behaviors, predict the intentions of traffic participants, and compensate for their errors, thus increasing safety and socio-acceptability.
3) ***Interaction with the driver:*** the personalization in the sense"to suit the automated task to the preferences and needs of the driver," we believe that it should consider the driver behavior that is understood as the intentional actions issued from the driver's inner mental thought and unintentional characteristics [71], but while answering the in which situation question. The driving task for an intelligent vehicle still needs the driver to remain active and engaged to take back the control [72], considering that the driver cognitive level is primordial for the system.

To achieve such a cross-cultural intelligent system, we are now facing the challenge of defining the value of society across different scenarios and translating the set of ethical rules into a language that the vehicle can understand independently from any human intervention. We need then to propose a cognitive architecture capable of being socially and culturally aware and in the sense of being able to abstract the situational information on the road, to retrieve what is relevant for its application. The framework should emulate the way expert drivers understand human interactions on the road and comply with mutually-accepted rules learned from countless experiences, we can do so by enabling the intelligent vehicle to memorize, reason, experiential update its knowledge, and extend the generalized knowledge learned to new scenes that were previously unknown and gain in adaptability and dynamic reconfiguration to face the environment changes and different Human/vehicles interactions [73]. We recognize that the challenges are enormous, adding to what was mentioned, the simplicity of most car simulators, especially the lack of realism when addressing the social and cultural aspects.

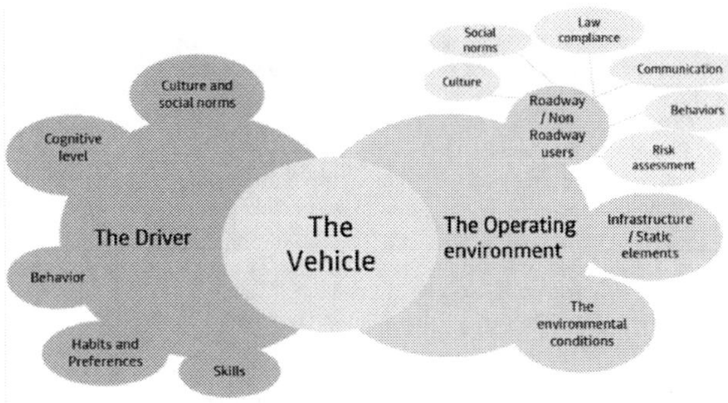

Fig. 3. Driver-vehicle-environmement in the loop.

VI. Conclusion And Future work

In this paper, we highlight the importance of considering seriously one of the most challenging problems facing the intelligent vehicle today, which is the cross-culture adaptability. We have taken a first step towards that by defining the culture with regards to driving and emphasizing the bias in the development of intelligent vehicles to date. We have provided a review of the current state of the art for personalization in advanced driver assistance systems and social autonomous driving. Our concern was the driver models used for personalization. The main objective of ADAS customization is to increase the system usability and, as a result, the driver acceptance. This is particularly crucial in applications safety-related such as Forward Collision Warning, where alarms and their timing should be tailored to the needs and skills of the driver to prevent the system underuse. The state of the art for social autonomous vehicles was then reviewed, including some studies that analyze interactions with other road and non-road users as well as anthropological and legal dilemmas in an urban environment. Finally, we discuss what a cross-culture adaptive intelligent vehicle could be considered. It should possess a model of its environment, allowing is to predict the intentions of road and non-road users, and thus, its safe driving is predetermined based on meeting the expectations of others. The intelligent vehicle should possess this set of knowledge, skills, and competencies to recognize, understand, and adapt to social and cultural differences. The challenges are enormous; defining the value of society across different scenarios and proposing a cognitive-based architecture for the system would be our next step.

References

[1] J. Moeckli, and J. D. Lee, "The making of driving cultures," Improving Traffic Safety Culture in the United States, vol. 38, no. 2, pp. 185-192, 2007.

[2] EuroNcap, 2020 assisted driving tests. Available from: https://www.euroncap.com/en/vehicle-safety/safety-campaigns/2020-assisted-driving-tests/ 2022-11-02.

[3] C. Ranasinghe et al., "Autonomous vehicle-pedestrian interaction across cultures: towards designing better external human machine interfaces (ehmis)," Extended Abstracts of the 2020 CHI Conference on Human Factors in Computing Systems, pp. 1–8, 2020.

[4] World Health Organization. Global status report on road safety 2018. Available from: https://www.who.int/publications/i/item/9789241565684/ 2022-11-01.

[5] T. Nordfjærn, S. Jørgensen, and T. Rundmo, "A cross-cultural comparison of road traffic risk perceptions, attitudes towards traffic safety and driver behaviour," Journal of Risk Research, vol. 14, no. 6, pp. 657-684, 2011.
[6] B. Pires, J. Owens and M. Jenkins, Walking and biking in an automated future series: part i: The promise and challenges of automated technologies. Available from: https://www.pedbikeinfo.org/webinars/ 2022-11-01.
[7] S. Mohammed, Seven life lessons from india's hazardous traffic chaos -how we manage? Available from: https://shahmm.medium.com/what-could-we-learn-from- indias-hazardous-traffic-chaos-9f5f849edd2f/ 2022-11-01.
[8] J. A. Mahmood, "What do car horns say? an overview of the non-verbal communication of horn honking," Open Journal of Social Sciences, vol. 9, no. 8, pp. 375-388, 2021.
[9] G. Tiwari, Walking in indian cities – a dailyagony for millions. Available from: https://www.thehinducentre.com/the-arena/current-issues/walking-in-indian-cities-a-daily- agony-for-millions/article65551959.ece/ 2022-11-01.
[10] A. Acharyya, India's traffic condition and tips to improve. Available from: https://medium.com/@ajoy.acharyya/indias-traffic-condition-and-tips-to-improve-3e1803c10938/ 2022-11-01.
[11] K. Othman, "Public acceptance and perception of autonomous vehicles: a comprehensive review," AI and Ethics, 2021, vol. 1, no. 3, pp. 355-387, 2021.
[12] W. Fastenmeier and H. Gstalter, "Driving task analysis as a tool in traffic safety research and practice," Safety Science, vol. 45, no. 9, pp. 952-979, 2007.
[13] E. H. Schein, What is culture. Newbury Park, CA: Sage, pp. 243-253, 1991.
[14] N. Eliasoph and P. Lichterman, "Culture in interaction," American journal of sociology, vol. 108, no. 4, pp. 735-794, 2003.
[15] A. Swidler, Talk of love: How culture matters. University of Chicago press, 2013.
[16] Y. Wang and Q. Xu, "A Filed Study of External HMI for Autonomous Vehicles When Interacting with Pedestrians," HCI in Mobility, Transport, and Automotive Systems. Automated Driving and In-Vehicle Experience Design: Second International Conference, MobiTAS 2020, Held as Part of the 22nd HCI International Conference, HCII 2020, Copenhagen, Denmark, Proceedings, Part I 22. Springer International Publishing, pp. 181-196, 2020.
[17] D. Moore, R. Currano, G. E. Strack, and D. Sirkin, "The case for implicit external human-machine interfaces for autonomous vehicles," Proceedings of the 11th international conference on automotive user interfaces and interactive vehicular applications, pp. 295-307, 2019.
[18] T. Özkan et al., "Cross-cultural differences in driving behaviours: A comparison of six countries," Transportation research part F: traffic psychology and behaviour, vol. 9, no. 3, pp. 227-242, 2006.
[19] World Health Organization, Regional Office for Europe. European regional status report on road safety 2019. Available from: https://www.who.int/europe/publications/i/item/9789289054980/ 2022-11-01.
[20] R. Factor, D. Mahalel, and G. Yair, "The social accident: A theoretical model and a research agenda for studying the influence of social and cultural characteristics on motor vehicle accidents," Accident Analysis & Prevention, vol. 39, no. 5, pp. 914-921, 2007.
[21] H. Singh and A. Kathuria, "Analyzing driver behavior under naturalistic driving conditions: A review," Accident Analysis & Prevention, vol. 150, pp. 105-908, 2021.
[22] L. Duan and F. Chen, "The future of advanced driving assistance system development in china," Proceedings of 2011 IEEE International Conference on Vehicular Electronics and Safety, pp. 238-243, 2011.
[23] A. Lindgren, F. Chen, P. W. Jordan, and H. Zhang, "Requirements for the design of advanced driver assistance systems-the differences between swedish and chinese drivers," International Journal of Design, vol. 2, no. 2, 2008.
[24] V. Butakov and P. Ioannou, "Personalized driver/vehicle lane change models for adas," IEEE Transactions on Vehicular Technology, vol. 64, no. 10, pp. 4422-4431, 2015.
[25] N. A. Stanton and M. S. Young, "A proposed psychological model of driving automation," Theoretical Issues in Ergonomics Science, vol. 1, no. 4, pp. 315-331, 2000.
[26] A. Gray, Y. Gao, J. K. Hedrick, and F. Borrelli, Robust predictive control for semi-autonomous vehicles with an uncertain driver model, 2013 IEEE Intelligent Vehicles Symposium (IV), Gold Coast, QLD, Australia, pp. 208-213, 2013.
[27] C. Miyajima, et al., "Driver modeling based on driving behavior and its evaluation in driver identification," Proceedings of the IEEE, vol. 95, no. 2, pp. 427-437, 2007.
[28] M. Plöchl and J. Edelmann, "Driver models in automobile dynamics application," Vehicle System Dynamics, vol. 45, no. 7-8, pp. 699-741, 2007.
[29] D. D. Salvucci, "Modeling driver behavior in a cognitive architecture," Human factors, vol. 48, no. 2, pp. 362-380, 2006.
[30] J. A. Michon, "A Critical View of Driver Behavior Models: What Do We Know, What Should We Do?," Human behavior and traffic safety, pp. 485-524, 1985.
[31] K. Brown, K. Driggs-Campbell, and M. J. Kochenderfer, A taxonomy and review of algorithms for modeling and predicting human driver behavior, arXiv preprint arXiv:2006.08832, 2020.
[32] D. M. Zaidel, "A modeling perspective on the culture of driving," Accident Analysis / Prevention, vol. 24, no. 6, pp. 585-597, 1992.
[33] E. Rich, "User modeling via stereotypes," Cognitive science, vol. 3, no. 4, pp. 329-354, 1979.
[34] Y. L. Murphey, R. Milton, and L. Kiliaris, Driver's style classification using jerk analysis, 2009 IEEE Workshop on Computational Intelligence in Vehicles and Vehicular Systems, Nashville, TN, USA, pp. 23-28, 2009.
[35] M. Jeon, B. N. Walker, and J. B. Yim, "Effects of specific emotions on subjective judgment, driving performance, and perceived workload," Transportation research part F: traffic psychology and behaviour, vol. 24, pp. 197-209, 2014.
[36] T. Lajunen, A. Corry, H. Summala, and L. Hartley, "Cross-cultural differences in drivers' self-assessments of their perceptual-motor and safety skills: Australians and finns," Personality and Individual differences, vol. 24, no. 4, pp. 539-550, 1998.
[37] P. C. Cacciabue and F. Saad, "Behavioural adaptations to driver support systems: a modelling and road safety perspective," Cognition, Technology Work, vol. 10, pp. 31-39, 2008.
[38] N. Oliver and A. P. Pentland, Graphical models for driver behavior recognition in a smartcar, Proceedings of the IEEE Intelligent Vehicles Symposium 2000 (Cat. No.00TH8511), Dearborn, MI, USA, pp. 7-12, 2000.
[39] B. Gao, K. Cai, T. Qu, Y. Hu, and H. Chen, "Personalized adaptive cruise control based on online driving style recognition technology and model predictive control," IEEE Transactions on Vehicular Technology, vol. 69, no. 11, pp. 12482-12496, 2020.
[40] A. Rosenfeld, Z. Bareket, C. V. Goldman, D. J. LeBlanc, and O. Tsimhoni, "Learning drivers' behavior to improve adaptive cruise control," Journal of Intelligent Transportation Systems, vol. 19, no. 1, pp. 18-31, 2015.
[41] J. Wang, L. Zhang, D. Zhang, and K. Li, "An adaptive longitudinal driving assistance system based on driver characteristics," IEEE Transactions on Intelligent Transportation Systems, vol. 14, no. 1, pp. 1-12, 2013.
[42] B. Zhu, S. Yan, J. Zhao, and W. Deng, "Personalized lane-change assistance system with driver behavior identification," IEEE Transactions on Vehicular Technology, vol. 67, no. 11, pp. 10293-10306, 2018.
[43] J. Wang, C. Yu, S. E. Li, and L. Wang, "A forward collision warning algorithm with adaptation to driver behaviors," IEEE Transactions on Intelligent Transportation Systems, vol. 17, no. 4, pp. 1157-1167, 2016.
[44] S. M. Iranmanesh, H. N. Mahjoub, H. Kazemi, and Y. P. Fallah, "An adaptive forward collision warning framework design based on driver distraction," IEEE Transactions on Intelligent Transportation Systems, vol. 19, no. 12, pp. 3925-3934, 2018.
[45] A. Abdelrahman, A. S. El-Wakeel, A. Noureldin, and H. S. Hassanein, "Crowdsensing-based personalized dynamic route planning for smart vehicles," IEEE Network, vol. 34, no. 3, pp. 216-223, 2020.
[46] M. Brackstone and M. McDonald, "Car-following: a historical review," Transportation Research Part F: Traffic Psychology and Behaviour, vol. 2, no. 4, pp. 181-196, 1999.
[47] Y. Feng, P. Iravani, and C. Brace, "A fuzzy logic-based approach for humanized driver modelling," Journal of advanced transportation, vol. 2021, pp. 1-13, 2021.
[48] M. Hasenj Ager, M. Heckmann, and H. Wersing, "A survey of personalization for advanced driver assistance systems," IEEE Transactions on Intelligent Vehicles, vol. 5, no. 2, pp. 335-344, 2020.

[49] M. Hasenj Ager and H. Wersing, "Personalization in advanced driver assistance systems and autonomous vehicles: A review," IEEE 20th International Conference on Intelligent Transportation Systems (ITSC), pp. 1–7, 2017.
[50] A. Ponomarev and A. Chernysheva, "Adaptation and personalization in driver assistance systems," 24th Conference of Open Innovations Association (FRUCT), Moscow, Russia, pp. 335-344, 2019.
[51] L. Halilaj, I. Dindorkar, J. Luettin, and S. Rothermel, "A Knowledge Graph-Based Approach for Situation Comprehension in Driving Scenarios," The Semantic Web: 18th International Conference, ESWC 2021, Virtual Event, June 6–10, 2021, Proceedings 18, Springer International Publishing, pp. 699–716, 2021.
[52] W. Schwarting, J. Alonso-Mora, and D. Rus, Planning and decision-making for autonomous vehicles, Annual Review of Control, Robotics, and Autonomous Systems, vol. 1, pp. 187-210, 2018.
[53] A. Rasouli and J. K. Tsotsos, "Autonomous vehicles that interact with pedestrians: A survey of theory and practice," IEEE Transactions on Intelligent Transportation Systems, vol. 21, no. 3, pp. 900-918, 2020.
[54] B. Jafary, E. Rabiei, M. Diaconeasa, H. Masoomi, L. Fiondella, and A. Mosleh, "A survey on autonomous vehicles interactions with human and other vehicles," 14th PSAM International Conference on Probabilistic Safety Assessment and Management, 2018.
[55] C. S. Fisk, "Game theory and transportation systems modelling," Transportation Research Part B: Methodological, vol. 18, pp. 301–313, 1984.
[56] P. Hang, C. Lv, C. Huang, J. Cai, Z. Hu, and Y. Xing, "An integrated framework of decision making and motion planning for autonomous vehicles considering social behaviors," IEEE Transactions on Vehicular Technology, vol. 69, no. 12, pp. 14458-14469, 2020.
[57] P. Hang, C. Lv, Y. Xing, C. Huang, and Z. Hu, "Human-like decision making for autonomous driving: A noncooperative game theoretic approach," IEEE Transactions on Intelligent Transportation Systems, vol. 22, no. 4, pp. 2076-2087, 2021.
[58] Y. Xia, Z. Qu, Z. Sun, and Z. Li, "A human-like model to understand surrounding vehicles' lane changing intentions for autonomous driving," IEEE Transactions on Vehicular Technology, vol. 70, no. 5, pp. 4178-4189, 2021.
[59] I. Kotseruba and J. K. Tsotsos, "Attention for vision-based assistive and automated driving: A review of algorithms and datasets," IEEE Transactions on Intelligent Transportation Systems, 2022.
[60] B. Škugor, J. Topić, J. Deur, V. Ivanović, and E. Tseng, "Analysis of a game theory-based model of vehicle-pedestrian interaction at uncontrolled crosswalks," In 2020 International Conference on Smart Systems and Technologies (SST), pp. 73–81, 2020.
[61] Y. Chen, C. Dong, P. Palanisamy, P. Mudalige, K. Muelling, and J. M. Dolan, "Attention-based hierarchical deep reinforcement learning for lane change behaviors in autonomous driving," 2019 IEEE/RSJ International Conference on Intelligent Robots and Systems (IROS), pp. 3697–3703, 2019.
[62] Z. Yang, Y. Zhang, J. Yu, J. Cai, and J. Luo, "End-to-end multi-modal multi-task vehicle control for self-driving cars with visual perceptions," 24th International Conference on Pattern Recognition (ICPR), pp. 2289–2294, 2018.
[63] S. Shalev-Shwartz, S. Shammah, and A. Shashua, Safe, multi-agent, reinforcement learning for autonomous driving, arXiv preprint arXiv:1610.03295, 2016.
[64] M. Toromanoff, E. Wirbel, F. Wilhelm, C. Vejarano, X. Perrotton, and F. Moutarde, "End to end vehicle lateral control using a single fisheye camera," In 2018 IEEE/RSJ International Conference on Intelligent Robots and Systems (IROS), pp. 3613–3619, 2018.
[65] M. Toromanoff, E. Wirbel, and F. Moutarde, "End-to-end model-free reinforcement learning for urban driving using implicit affordances," 2020 IEEE/CVF Conference on Computer Vision and Pattern Recognition (CVPR), pp. 7151–7160, 2020.
[66] P. Wang, C. Y. Chan, and A. de La Fortelle, A reinforcement learning based approach for automated lane change maneuvers, 2018 IEEE Intelligent Vehicles Symposium (IV), pp. 1379–1384, 2018.
[67] S. H. Lee and S. W. Seo, "A learning-based framework for handling dilemmas in urban automated driving," 2017 IEEE International Conference on Robotics and Automation (ICRA), pp. 1436–1442, 2017.
[68] J. Zhang and E. Ohn-Bar, "Learning by watching," Proceedings of the IEEE/CVF Conference on Computer Vision and Pattern Recognition, pp. 12711-12721, 2021.
[69] A. Bera, T. Randhavane, A. Wang, D. Manocha, E. Kubin, and K. Gray, "Classifying group emotions for socially-aware autonomous vehicle navigation," 2018 IEEE/CVF Conference on Computer Vision and Pattern Recognition Workshops (CVPRW), pp. 1152–11528, 2018.
[70] S. Chen, Z. Jian, Y. Huang, Y. Chen, Z. Zhuoli, and N. Zheng, "Autonomous driving: cognitive construction and situation understanding," Science China Information Sciences, vol. 62, pp. 1-27, 2019.
[71] Y. Xing, C. Lv, and D. Cao, "Driver behavior recognition in driver intention inference systems," Adv. Driv. Intent. Inference, vol. 258, pp. 99-134, 2020.
[72] M. Miyaji, H. Kawanaka and K. Oguri, "Driver's cognitive distraction detection using physiological features by the adaboost," 2009 12th International IEEE Conference on Intelligent Transportation Systems, St. Louis, MO, USA, pp. 1-6, 2009.
[73] E. Coelingh, P. Chaumette, and M. Andersson, Open-interface definitions for automotive systems application to a brake by wire system, SAE Transactions, pp. 151–159, 2002.

Federated Monocular 3D Object Detection for Autonomous Driving

Fangyuan Chi, Yixiao Wang, Panos Nasiopoulos, Victor C.M. Leung, Mahsa T. Pourazad
Department of Electrical and Computer Engineering
The University of British Columbia
Vancouver, BC, Canada
{fangchi, yixiaow, panosn, vleung, pourazad}@ece.ubc.ca

***Abstract*—In this paper, we propose and implement a novel method for 3D object detection in autonomous driving by applying federated mechanism to a monocular camera-based network. Our approach has several advantages over traditional 3D object detection methods that rely on LiDAR or other sensors, as it is more cost-effective and can be more easily integrated into existing autonomous driving systems. We use a federated learning framework, which allows us to train the model on a large amount of data covering a variety of scenarios without having to share the raw data with a central server. This allows us to reduce transmission bandwidth requirements and preserve the privacy of the data contributors, while still achieving high accuracy in 3D object detection. In our experiments, we evaluate our method on a variety of challenging real-world driving scenarios and show that it is able to accurately detect objects in 3D from a monocular camera view. Our results demonstrate the effectiveness of our approach and show its potential for use in autonomous driving systems.**

Keywords-monocular, 3D object detection, federated learning, autonomous driving.

I. INTRODUCTION

Safety remains the primary concern when people talk about autonomous driving. Autonomous vehicles are empowered by various Deep Learning (DL) models (i.e., perception, tracking, prediction, etc.), but these models are trained with either simulated data or controlled driving scenarios, with most autonomous vehicles still being tested in enclosed facility environments. It is, therefore, difficult to evaluate how they will perform in real-world driving scenarios, especially when the controlled environment is coupled with numerous unpredictable corners, emergencies, and occlusions. Human drivers gain experience over time, first watching the parents or others driving, then through driving schools and, finally, driving and improving their skills every year. Autonomous vehicles should do the same. Besides training deep learning models that enable autonomous driving in labs, autonomous vehicles should continuously learn from different driving scenarios of their own or others experience. Federated Learning (FL) is a promising approach that may address this problem. Instead of having autonomous vehicles to upload their perception data to the cloud to perform centralized training, as shown in Figure 1(a), FL allows autonomous vehicles to first train their local models with local collected data and share with each other their own experience through their DL models instead of sharing collected data, as

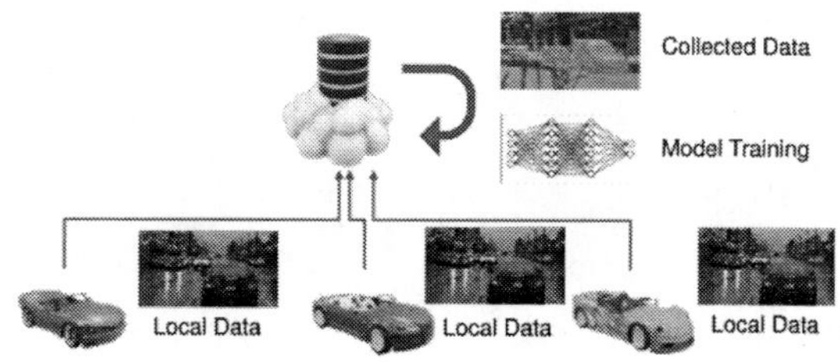

(a) Centralized Monocular Approach

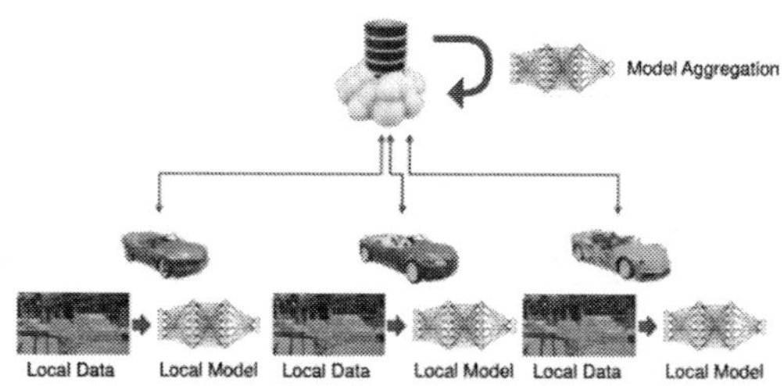

(b) Federated Monocular Approach

Figure 1. (a) Centralized approach which upload the local images from all vehicles to the central server, then train and release the global model, (b) Federated approach, in which each vehicle trains their own model with the local data, then the local models will be uploaded, aggreated and released.

shown in Figure 1(b). Recent works investigate if autonomous driving could benefit from FL. Some of them [1][2] are designing system architectures to ensure the efficiency of when and which vehicles should participate in the federation process. Other approaches are trying to address the problem that data collected locally are non-Independent and Identically Distributed (non-IID) [3][4]. Although existing methods demonstrate that FL could improve the accuracy in object detection, they have only been evaluated with 2D object detection [5] or 3D object detection using LiDAR [6]. However, it is not sufficient to use these two types of sensor data when evaluating methods for autonomous driving. More specifically, 2D object detection cannot output depth information or is hard to predict the distance of target objects, while LiDAR is expensive and is not commonly supported by autonomous vehicles, with the industry recently following a vision-based trend for autonomous driving [7][8]. In this paper, we investigate and verify that the performance of 3D object detection could benefit from leveraging federated learning with 3D image data collected by monocular cameras.

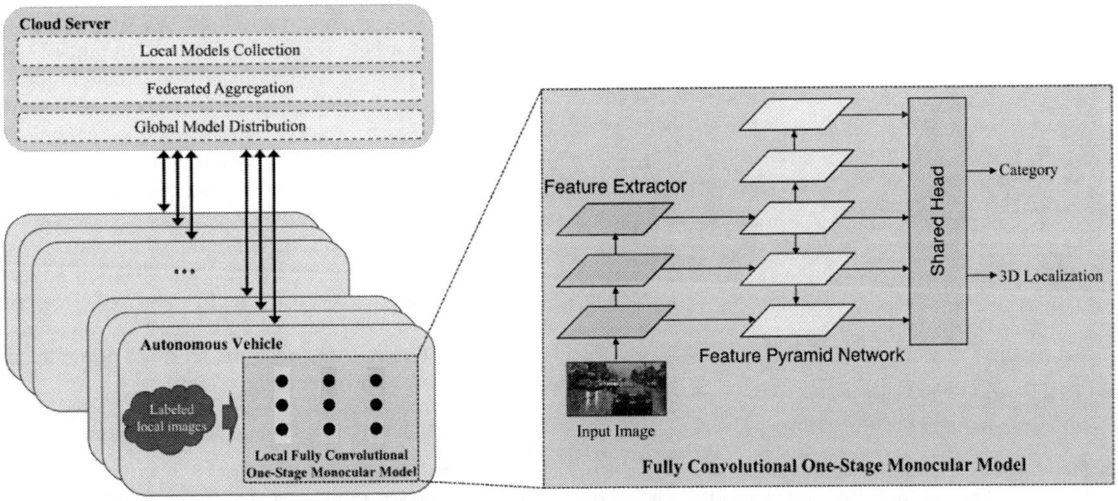

Figure 2. Overall architecture of the federated monocular 3D object detection approach.

The rest of the paper is organized as follows. Section II gives an overview of related work, while Section III presents our proposed approach. In Section IV, we evaluate and analyze the results. Section V concludes the paper.

II. OVERVIEW OF RELATED WORK

A. 3D Object Detection in Autonomous Driving

Object detection is a popular research topic in autonomous driving. Different approaches are proposed leveraging different kinds of sensors. For example, in [9] the RGB image captured by camera and point cloud obtained by LiDAR are fused together for 3D object detection. Evaluation results show that the detection accuracy increases since data from different modalities provide complimentary features (i.e., images provide semantic information while point cloud provides depth information to construct 3D surroundings). In [10], data obtained from LiDAR are used for 3D object detection. This approach considers long-range interactions among detection candidates. In [11], RGB-D images captured by monocular camera are used for 3D object detection. This vision-based approach is simpler, cost-efficient, and more practical compared to multi-modality-based approaches.

B. Federated Learning in Autonomous Driving

Federated learning is a machine learning approach that allows multiple participants to train a shared model without sharing their raw data. This is particularly useful in the context of autonomous driving, where data from individual vehicles may be sensitive or proprietary. With federated learning, each vehicle can train a local model on its own data, and then share the model updates with a central server. The server can then aggregate the updates and use them to improve a shared global model, without ever having access to the raw data. This approach has several potential benefits for autonomous driving. For example, it allows vehicles to learn from each other without sharing sensitive data, and can enable the development of more robust and accurate models by leveraging data from a larger and more diverse set of vehicles [1]-[6]. Additionally, federated learning can enable real-time updates to the global model, allowing vehicles to quickly adapt to changing conditions and improve their performance over time. Overall, federated learning has the potential to play a significant role in the development of autonomous driving systems.

III. OUR PROPOSED METHOD

In this paper, we propose a federated monocular 3D object detection approach for autonomous driving. The overall architecture of our method is illustrated in Figure 2. The left part of the figure shows the distributed federated learning mechanism, while the right part shows the local monocular model on each vehicle, which is trained to make predictions for 3D object detection.

A. Federated Learning-based Collaboration

The federated learning mechanism adopted in our approach includes the following 3 steps. 1) First, each vehicle trains a local monocular model on its own. This allows it to learn from its own data without sharing it with other vehicles or a central server. 2) After training the local models, each vehicle shares the model updates (e.g., weights, biases) with the central server. The server can then aggregate all these updates and use them to improve the global model. 3) Once the global model has been updated, the central server distributes the updated model to all the vehicles. Each vehicle can then use the updated global model to improve its own local model and continue the training process. This 3-step process can be repeated as necessary to continue improving the performance of the global model and enable vehicles to learn from each other. Over time, the global model should become more accurate and robust, allowing vehicles to make better decisions and improve their individual performance in real world conditions. Note that, during the federated training

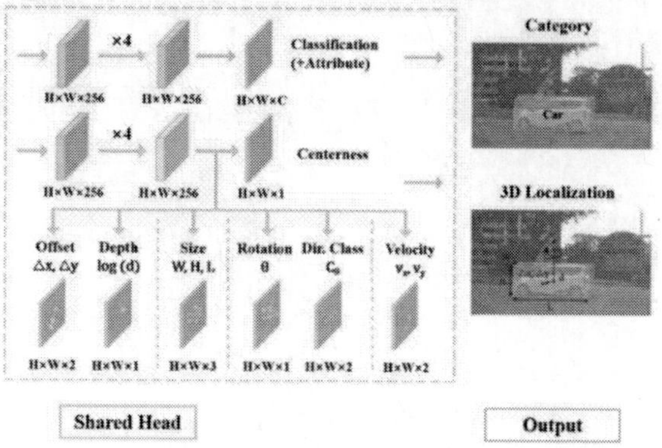

Figure 3. Illustration of the shared head in the local monocular 3D object detection model.

process, the images captured on each vehicle are not sent to the central server, keeping possible sensitive information on the local vehicle, thus eliminating privacy concern issues.

B. *Local Monocular 3D Object Detection Model*

We employ a fully convolutional one-stage monocular 3D object detection model for detecting objects in 3D from a single camera view [11]. This is an object detection approach that uses a Convolutional Neural Network (CNN) to predict the 3D bounding boxes and class probabilities of objects in an image.

One key advantage of this approach is that it is fully convolutional, meaning that the CNN can operate on input images of any size, and produce output predictions for each pixel in the image. This allows the model to be used on images of varying resolutions, without the need for manual resizing or cropping. Additionally, this approach is a one-stage method, meaning that it uses a single neural network to make all of its predictions. This makes the model more efficient and faster to run, as it does not require multiple stages of processing or separate networks for different tasks.

In the local monocular 3D object detection model, the ResNet101 [12] is employed as the feature extractor and the Feature Pyramid Network (FPN) [13] as the neck. The ResNet101 network is a well-known and widely used architecture for image classification and object detection tasks. It is a deep CNN that is composed of multiple convolutional layers, residual blocks, and pooling layers, and is designed to be highly efficient and accurate. By using ResNet101 as the feature extractor in our model, we can take advantage of its proven performance and efficiency, and extract high-quality features from the input images. The FPN is a network architecture that is commonly used in object detection tasks to improve the model's ability to detect objects at different scales. It is composed of a pyramid of feature maps, with each level of the pyramid representing features at a different scale. By using FPN as the neck in our model, we can improve the model's ability to detect objects such as pedestrians and cars at different distances from the camera.

The FPN neck is followed by a shared head, shown in Figure 3. Using a shared head to output the class of objects and 3D bounding boxes in a fully convolutional one-stage monocular 3D object detection model can have several benefits. First, a shared head allows the model to make predictions for both the class of objects and the 3D bounding box in a single pass, which can make the model more efficient and faster to run. This is particularly useful in real-time applications such as autonomous driving, where it is important to make predictions quickly and accurately. Second, a shared head can improve the model's overall performance, as it allows the CNN to learn features that are relevant for both tasks simultaneously. This not only can lead to more accurate predictions but also to better generalization to new data. Finally, a shared head can simplify the model's architecture, making it easier to train and optimize. This can save time and resources, and can make the model more portable and easier to integrate into different applications.

IV. EXPERIMENTAL RESULTS AND DISCUSSION

We used the nuScenes dataset to evaluate the performance of our federated monocular 3D object detection method. The nuScenes dataset is a large-scale dataset of annotated images and point clouds captured in real-world driving scenarios [14]. It contains a rich variety of data, including different environments, weather conditions, and vehicle types, making it an ideal testbed for evaluating our method. To perform our evaluation, we split the nuScenes dataset into 10 parts, each representing data from a different vehicle.

We then train our local monocular models on each of these vehicle datasets (10% of the whole dataset), both with and without federated learning. This allows us to compare the performance of our method with and without federated learning. We train all the models for 12 epochs at a batch size of 4 on a NVIDIA Tesla V100L graphic card. The learning rate is set to $5e^{-3}$ and is halved in both the 8th and 11th epoch. For our approach, the trained model weights from the 10 vehicles are aggregated in an average manner in every epoch.

After training our federated monocular 3D object detection model, we use the mean Average Precision (mAP) and NuScenes Detection Score (NDS) metrics to evaluate its performance. These metrics are commonly used to evaluate object detection algorithms, and allow us to compare our method to other state-of-the-art approaches. The mAP metric measures the average precision of the model across all classes and all thresholds. It is calculated by averaging the precision of the model at different recall levels, and is a useful metric for comparing the overall performance of different object detection models. The NDS metric is specific to the nuScenes dataset, and measures the overall performance of the model in terms of both precision and recall. It is calculated as the harmonic mean of the average precision and average recall of the model, and is a useful metric for evaluating object detection models on the nuScenes dataset.

Figure 4 (a) shows the performance comparison of our method with and without federated learning on the nuScenes dataset in NDS against epochs, while Figure 4 (b) shows the same comparison in mAP vs epochs. We observe that the NDS of our method is significantly higher when using federated learning, reaching 41.18%. This demonstrates that our method is able to improve both the precision and recall of its predictions, resulting in more complete and accurate

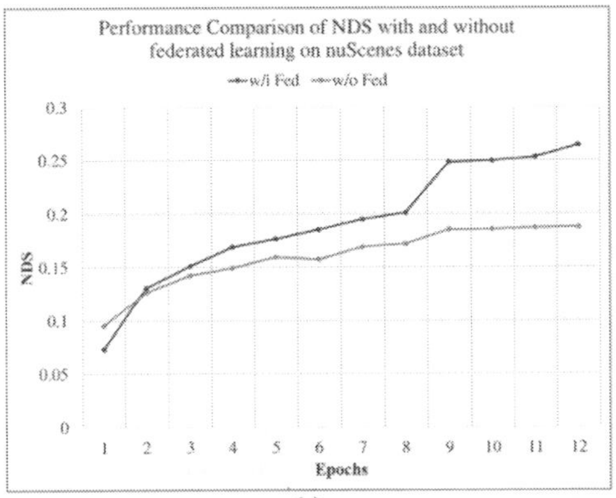

(a)

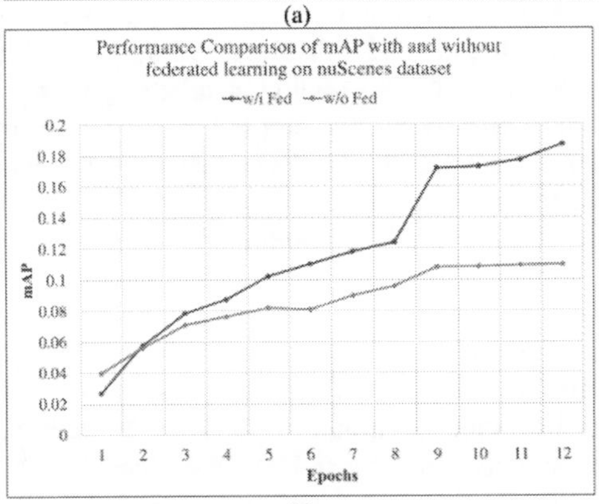

(b)

Figure 4. The validation results comparison at each training epoch on two metrics, (a) NDS, (b) mAP

detections of objects in the scene. Similarly, we observe that mAP of our method is 70% higher when using federated learning, indicating that our method is able to make more accurate predictions when using data from multiple vehicles, as opposed to training only on local data from a single vehicle. The performance is summarized in Table I.

Overall, our evaluation results show that the adopted federated learning mechanism is able to significantly improve the prediction performance of our federated monocular 3D object detection approach, leading to higher mAP and NDS scores compared to training only with local data. This demonstrates the effectiveness of our method and its potential for use in autonomous driving systems.

V. CONCLUSION

In conclusion, this paper has presented a novel method for 3D object detection in autonomous driving using only a monocular camera. Our approach uses federated learning to train a deep neural network that is able to detect objects in 3D from a single camera view, and has several advantages over traditional methods that rely on LiDAR or other sensors. We have evaluated our method on a variety of challenging real-world driving scenarios and showed that it is able to accurately detect objects in 3D from a monocular camera view. These results demonstrate the effectiveness of our approach and suggest its potential for use in autonomous driving systems.

TABLE I. COMPARISON OF NDS AND MAP METRICS BETWEEN THE BASELINE AND OUR PROPOSED FEDERATED APPROACH

Method	Backbone	Data Ratio	NDS	mAP
Baseline	ResNet101	10%	0.187	0.110
Ours	ResNet101	10%	**0.264 (+41.18%)**	**0.187 (+70%)**

REFERENCES

[1] A. Nguyen et al., "Deep Federated Learning for Autonomous Driving," 2022 IEEE Intelligent Vehicles Symposium (IV), 2022, pp. 1824-1830.

[2] S. Wang et al. Federated Deep Learning Meets Autonomous Vehicle Perception: Design and Verification[J]. arXiv preprint arXiv:2206.01748, 2022

[3] Q. Li, B. He, and D. Song, "Model-contrastive federated learning" Proc. IEEE/CVF Conference on Computer Vision and Pattern Recognition. 2021, pp. 10713-10722.

[4] T. Zeng, O. Semiariy, M. Chen, W. Saad and M. Bennis, "Federated Learning on the Road Autonomous Controller Design for Connected and Autonomous Vehicles," IEEE Transactions on Wireless Communications, 2022, vol. 21, no. 12, pp. 10407-10423

[5] D. Jallepalli, N. C. Ravikumar, P. V. Badarinath, S. Uchil and M. A. Suresh, "Federated Learning for Object Detection in Autonomous Vehicles," IEEE Seventh International Conference on Big Data Computing Service and Applications (BigDataService), 2021, pp. 107-114.

[6] B. Luca, S. Stefano, B. Mattia, and N. Monica, "Decentralized federated learning for extended sensing in 6G connected vehicles", Vehicular Communications, 2022, vol. 33, pp.100396, ISSN 2214-2096

[7] R. Trabelsi, R. Khemmar, B. Decoux, J. Y. Ertaud, and R. Butteau, "Recent advances in vision-based on-road behaviors understanding: a critical survey". Sensors, 2022, vol. 22, no. 7, pp. 2654

[8] E. Zablocki, H. Ben-Younes, P. Pérez, and M. Cord. "Explainability of deep vision-based autonomous driving systems: Review and challenges". International Journal of Computer Vision, 2022, vol. 130, no. 10, pp. 2425-2452.

[9] X. Chen, H. Ma, J. Wan, B. Li, and T. Xia, "Multi-view 3D Object Detection Network for Autonomous Driving," IEEE Conference on Computer Vision and Pattern Recognition (CVPR), Honolulu, HI, Jul. 2017, pp. 6526–6534.

[10] C. He, H. Zeng, J. Huang, X. Hua, and L. Zhang, "Structure aware single-stage 3d object detection from point cloud" Proc. IEEE/CVF conference on computer vision and pattern recognition. 2020, pp. 11873-11882.

[11] T. Wang, X. Zhu, J. Pang, and L. Dahua, "Fcos3d: Fully convolutional one-stage monocular 3d object detection" in Proceedings of the IEEE/CVF International Conference on Computer Vision. 2021, pp. 913-922.

[12] K. He, X. Zhang, S. Ren, and J. Sun, "Deep Residual Learning for Image Recognition", IEEE Conference on Computer Vision and Pattern Recognition (CVPR), Las Vegas, NV, USA, Jun. 2016, pp. 770–778.

[13] T. Y. Lin, P. Dollar, R. Girshick, K. He, B. Hariharan, and S. Belongie, "Feature Pyramid Networks for Object Detection", IEEE Conference on Computer Vision and Pattern Recognition (CVPR), Honolulu, HI, Jul. 2017, pp. 936–944.

[14] H. Caesar et al., 'nuScenes: A Multimodal Dataset for Autonomous Driving', IEEE/CVF Conference on Computer Vision and Pattern Recognition (CVPR), Seattle, WA, USA, Jun. 2020, pp. 11618–11628.

Simulation Environment for Validation of Automated Lane-Keeping System

Jiri Vlasak*†, Michal Sojka*, Zdeněk Hanzálek*
*Czech Institute of Informatics, Robotics and Cybernetics, †Faculty of Electrical Engineering,
Czech Technical University in Prague
Jugoslávských partyzánů 1580/3, 160 00 Praha 6, Czech Republic
{jiri.vlasak.2,michal.sojka,zdenek.hanzalek}@cvut.cz

***Abstract*—Automated Driving Systems (ADS) need to be validated in a wide range of conditions to ensure the safety of their operation. It is impossible to validate everything in a real environment, and simulation is the only viable alternative to cover testing under all needed conditions. We present the architecture of a simulation environment based on the CARLA simulator, aimed at validating the Automated Lane-Keeping System (ALKS), the first ADS with available legislation for its approval. We propose to simplify the development and deployment of such a complex software simulation environment through the use of the Nix package manager. We also propose how the example scenarios distributed with CARLA can be extended to make them suitable for validation of ALKS.**

***Index Terms*—Automated Driving System; CARLA simulator; Robot Operating System; Automated Lane-Keeping System.**

I. Introduction

The development of Automated Driving Systems (ADS) requires a lot of validation and testing activities to ensure safe operation. Automated Lane Keeping System (ALKS) is the first ADS for which there is a legal document specifying the requirements for its approval: United Nations (UN) Regulation No. 157 [1]. Compared to similar regulations issued in the past, this document contains only general requirements without specific detailed instructions on what and how to test. It is up to the approving organization to develop precise procedures that will enable it to assess the safety of the ALKS implementation.

To this end, many approval bodies are preparing procedures and technical equipment for the approval process, and similarly, car makers are working on their internal validation and testing procedures. This work is a first step toward the same goal, but on a smaller scale. As an academic institution, we are developing an ALKS-like function to drive a real car under a set of limiting conditions. In particular, we limit the functionality to a subset of weather conditions and omit some required functionality like detection of approaching emergency vehicles. At the same time as the ALKS function, we are developing a simulation environment allowing to test out the ALKS implementation in a virtual environment before deploying it in the real vehicle. The goal is to close the loop between development and validation, gain experience, and provide feedback to other organizations working with real ALKS implementations.

In this paper, we outline the architecture of our initial simulation environment, which is based on open-source software, namely the CARLA simulator. Since the software environment for simulation and validation integrates software components from different sources, version conflicts between different components often occur. We propose to solve these integration problems using the Nix package manager. We publish our initial implementation for use by others. Additionally, we analyze which ALKS validation scenarios, as required by current legislation, can be reused from CARLA and related tools and how they need to be extended for ALKS validation.

This paper is structured as follows. We review related work in Section II. Then, in Section III we describe the architecture of our simulation environment followed by the analysis of simulation scenarios in Section IV. We conclude in Section V.

II. Related work

Based on the definitions and taxonomy from SAE International [2], this paper targets verification of ALKS, in the scenario-based simulation environment.

Requirements for the ALKS are given by UN Regulation No. 157 [1] (regulation in short). The regulation introduces the required behavior of the ALKS (Dynamic Driving Task – DDT) within predefined conditions (Operational Design Domain – ODD). However, the regulation lacks exact parameters for the test scenarios.

To overcome the problem of missing parameters, Tenbrock et al. [3] present a methodology for finding relevant scenarios in real-world data and extracting the scenarios parameters. They apply their methodology to the highD dataset [4] extracting more than 340 scenarios into OpenSCENARIO format to be later used in CARLA or esmini simulators. They named the extracted dataset ConScenD.

In this work, we use CARLA [5] – an open-source simulator for driving automation systems testing – along with the ScenarioRunner [6] to execute the scenarios. The CARLA simulator already includes basic scenarios like lane-keeping, vehicle following, or cut-in and cut-out situations. We aim at extending these scenarios to cover all parameters from the regulation.

Riedmaier et al. [7] present a state of the art survey on scenario-based approaches to safety assessments of driving automation systems. They discuss two approaches for scenario

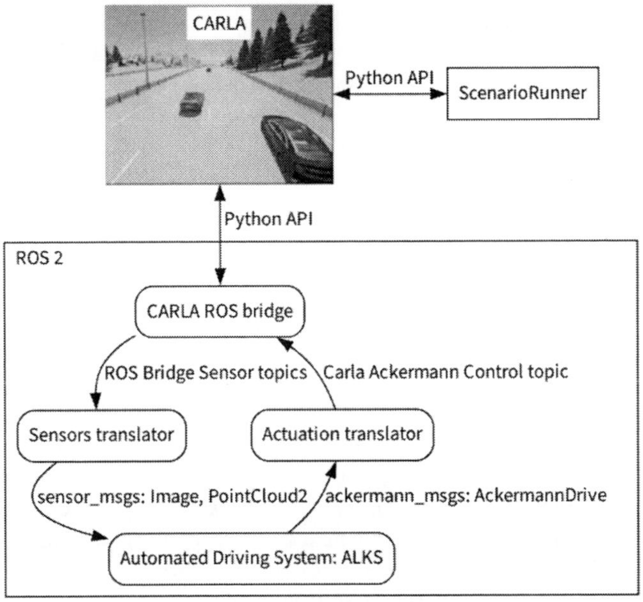

(a) ALKS in simulation environment.

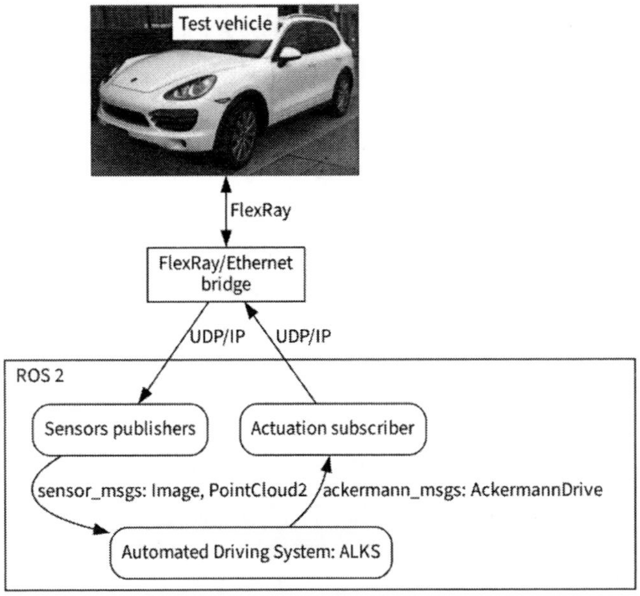

(b) ALKS in test vehicle.

Fig. 1. Interfacing of Automated Driving System into simulated and real environments.

generation: knowledge-based scenario generation and data-driven scenario extraction. By this distinction, basic scenarios included in CARLA area knowledge-based, and ConScenD scenarios are data-driven. The authors also discuss the difference between testing- and falsification-based scenario selection. The former approach aims at a subset of scenarios and generalize the results. The latter aims at finding a violation of the safety requirements. Furthermore, the authors identify formal verification as an alternative to scenario-based testing. They propose formal verification for the planning module and the scenario-based testing for the whole system.

Weissensteiner et al. [8] introduce a simulation framework for scenario-based virtual validation of ALKS with its necessary subsystems, including the interfaces between these subsystems. They test the simulation framework on ALKS in two Operational Design Domains using AVL Model.CONNECT, CarMaker, and CARLA.

III. Simulation environment

In this section, we describe the architecture of our simulation environment (Figure 1a). It is based on an open source simulator CARLA [5] and the goal is to test the unmodified implementation of ALKS software developed for a real vehicle (Figure 1b).

A. Architecture

The simulation environment depicted in Figure 1a uses the CARLA simulator as the main simulation engine and its module ScenarioRunner. ScenarioRunner is a CARLA client application written in Python and we use it to initialize the scenario and to control the vehicles (other than the ego vehicle) in the scenario.

We develop the ALKS in the Robot Operating System (ROS) [9] so we also use *CARLA ROS bridge* [10] as an interface between ROS and CARLA. CARLA ROS bridge is a ROS package providing the so called ROS node that acts as a CARLA client and translates data from CARLA to ROS and vice versa. In the ROS terminology, it acts as a ROS publisher for data from CARLA and as a ROS subscriber for data that go in the other direction. CARLA ROS bridge is accompanied by related ROS packages as CARLA Ackermann control and CARLA spawn objects. CARLA Ackermann control allows controlling the ego vehicle in the simulation via well known AckermannDrive ROS messages. CARLA spawn objects is used to manage vehicles simulated in CARLA from ROS. Particularly, we use it to spawn the ego vehicle.

The CARLA ROS bridge publishes data from simulated sensors in CARLA-specific messages. To use them with the ALKS implementation for the test vehicle, we convert them to the format expected by the implementation. This is implemented in the *Sensors translator* ROS node. Similarly, the messages in the other direction are converted from vehicle-specific to CARLA-specific format by the *Actuation translator* ROS node.

The Automated Driving System node in Figure 1a implements the Automated Lane-Keeping System (ALKS). It is a collection of multiple cooperating ROS nodes with specific purposes such as perception, maneuver decision, trajectory planning, and trajectory execution. The internal architecture of the *Automated Driving System* block is out of the scope of this paper.

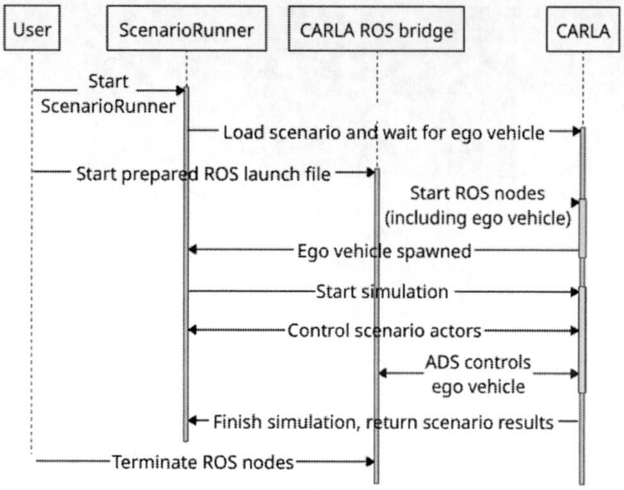

Fig. 2. Sequence diagram of running scenarios in the simulation environment.

B. Execution

To execute the simulations, we run the individual components as shown in the sequence diagram in Figure 2. Currently, we run them manually. Automation is planned for the future.

First, we start the ScenarioRunner with arguments telling it how to connect to the CARLA simulator, which scenario to load, and that the ScenarioRunner should not control the ego vehicle. After the ScenarioRunner prepares the scenario and waits for the ego vehicle, we start all ROS nodes from Figure 1a via a ROS launch file. This step includes spawing the ego vehicle itself. After all required ROS nodes are running and the ego vehicle is spawned, the simulation starts automatically. Individual simulation time-steps are triggered by the ScenarioRunner. ScenarioRunner also controls the movement of simulation vehicles except the ego vehicle, which is controlled by the ALKS ROS node, as shown in Figure 1a. When end conditions of the scenario are satisfied, the simulation ends, and the ScenarioRunner finishes. Then, we manually stop the remaining ROS nodes.

C. Reproducible development environment

The testing environment presented above allows to simulate selected scenarios for ALKS validation. However, setting up the environment on one's computer requires significant effort. For Linux, which we target for our development, CARLA is officially supported only on Ubuntu 18.04. Using it with newer Ubuntu versions works with a few undocumented tweaks. Using it with non-Ubuntu distribution is more difficult. For advanced use of CARLA, such as adding new maps or vehicle models, building CARLA from source is required. This is even more complex, as it needs installing packages from unofficial sources, easily leading to broken systems. At least, this is our experience with students trying to build CARLA themselves. Building CARLA from its source code requires about 130 GB of disk space and hours of compilation time. Any error during the build process multiplies the needed time.

Some of the problems related to building and installing CARLA can be mitigated by using Docker and pre-built CARLA images. However, using Docker brings other challenges that need to be overcome such as making the GPU (Graphics Processing Unit) available inside the Docker containers.

While the above mentioned problems can be resolved with some effort, we argue that the effort is better spent elsewhere than repeatedly trying to reproduce commands from CARLA documentation. Therefore, we propose to use the Nix package manager [11] to manage the software stack for ADS testing and validation. Nix revolutionizes the software building and deployment process by providing strictly controlled environment for software builds, which makes it easy to build complex software stacks reproducibly, i.e., results are bit-by-bit equivalent, in a way that works the same on any Linux distribution. This is achieved by the following features of Nix: (1) Every build command can access only explicitly specified dependencies, the rest of the system is "invisible" to it and (2) any piece of software, called *store object* in Nix terminology, is identified by a hash of all inputs (dependencies) and the commands used to build it. Going into details is out of scope of this paper, but an interested reader can refer to [12], which describes similar problems that we experienced with CARLA, explains how Nix helps to solve them, and why is the Nix solution better than using Docker. The main problem of building CARLA "the Nix way" is the fact that CARLA, and the Unreal Engine, which CARLA is based on, try to achieve reproducible build by its own imperfect way, i.e., by using a custom build system and by downloading prebuilt versions of some, but not all dependencies.

The result of our work-in-progress effort is available in a GitHub repository [13]. Currently, it provides two main functionalities:

- An environment for building CARLA from source. Such an environment contains all the needed dependencies like libraries and compilers in correct versions.
- CARLA client libraries and Python bindings packaged as Nix expressions, allowing their compilation and use on any Linux distribution. This can be used to develop CARLA clients even if binary CARLA packages are unavailable, for example due to the fact that your distribution provides only newer Python versions than those required by CARLA binary packages.

The instructions for how to use the repository are provided in its README file.

IV. Validation scenarios

The UN Regulation No. 157 [1] defines in Annex 5 so called test scenarios, which should be used to validate ALKS implementations. We are interested in simulating those scenarios during ALKS development before testing them in real world.

The ScenarioRunner for CARLA comes with several example scenarios. Here, we analyze which of those example scenarios can be used for validation of ALKS and how they need to be modified or extended for ALKS testing. The

TABLE I. ALKS TEST SCENARIOS AND HOW THEY MAP TO CARLA SCENARIORUNNER EXAMPLES.

ALKS Test Scenario	Similar ScenarioRunner Examples	Extensions	Notes
1. Lane Keeping	FollowLeadingVehicle{5,7}	§ 1 to 5	§ 10
2. Avoiding a collision with a road user or object blocking the lane	StationaryObjectCrossing5, DynamicObjectCrossing5	§ 3 and 6	
3. Following a lead vehicle	FollowLeadingVehicle{5,7}	§ 1 to 5	
4. Lane change of another vehicle into lane	CutInFrom_left_Lane, CutInFrom_right_Lane	§ 7	
5. Stationary obstacles after lane change of the lead vehicle	ChangeLane1	§ 3 and 6	
6. Field of View test	N/A	§ 8	§ 11
7. Lane changing	OtherLeadingVehicle{1,2,4,5,6}	§ 9	§ 12
8. Avoiding emergency manoeuvre before a passable object in the lane	FollowLeadingVehicle{5,7}	§ 1	

(§1) Use different vehicle types of other vehicles, i.e., leading, cutting-in, in the scenario. Namely use: passenger car, powered two-wheeler (PTW), and other vehicle. (§2) Test for different the lead vehicle speeds of (constant, realistic speed profile, braking) and different steering (still, swerving, different lateral positions in lane). (§3) Test different roads segments (straight, various curvatures). (§4) Make scenario timeout longer (5 minutes) and update end conditions. (§5) Extend these scenarios with another vehicle driving close but in an adjacent lane. (§6) Test multiple stationary objects (passenger car, PTW, pedestrian, partially or fully blocked lane, multiple obstacles). (§7) Parameterize scenario with different Times to Collision (TTC), distances, and relative velocities to test scenarios where collision can be avoided as well as scenarios where collision is unavoidable. This includes different longitudinal speed (constant, accelerating, decelerating) and different lateral velocity (constant, accelerating, decelerating) of cutting-in vehicle. Examples of simulations when ALKS should avoid the collision for cut-in, cut-out, and deceleration scenarios are depicted in Appendix 1 of Annex 5 of the UN Regulation No. 157 [1]. (§8) New test scenarios will need to be created. These should contain stationary objects (pedestrian, PTW) on the outer edge of adjacent lanes and within the ego lane and PTW approaching from different directions to the ego vehicle.
(§9) Test different situations when Lane Change Maneuver (LCM) is either possible or impossible due to other vehicles (passenger car, PTW) approaching from different sides of the ego vehicle. (§10) To test the functionality of cruise control without a leading vehicle, we need to create a new scenario because the ScenarioRunner cruise control examples are not positioned on the highway. (§11) These tests target real vehicle sensors. In simulations, properties of simulated sensors can be configured to be arbitrarily good or bad. (§12) Only for ALKS implementations capable of lane change procedure (LCP).

results are summarized in Table I, where each line represents one ALKS test scenario and which of the ScenarioRunner examples are most suitable for simulating it (if any). Each ScenarioRunner example has a name, e.g., *FollowLeadingVehicle* and can be parameterized by several parameters. The particular set of parameters is identified by the number after the scenario name; the values of the parameters are specified in the XML file of the example. If multiple parameter sets are suitable, we denote it with their numbers in curly braces, e.g., *FollowLeadingVehicle{5,7}*.

In addition to ALKS test scenarios from the UN Regulation 157, many other scenarios will need to be created to test full functionality of the ALKS implementation. These include, e.g., scenarios for testing the implementation of the minimum risk maneuver, which should stops the vehicle in a safe way if the driver is not responding to the request to take over vehicle control.

V. CONCLUSION

In this paper, we present an initial version of the simulation environment to validate an implementation of the Automated Lane-Keeping System developed for a real vehicle using the Robot Operating System. The simulation environment is based on the CARLA simulator.

Building the CARLA simulator from source code to add custom vehicles or sensors is a time-consuming task. To avoid this effort, we propose to use the Nix package manager and publicly share the repository with our ongoing work.

Furthermore, we analyzed which CARLA example scenarios are suitable for implementing tests specified in Annex 5 of UN Regulation No. 157 [1].

The goal of our future work is to complete Nix packaging of the CARLA simulator with related tools and develop ALKS validation scenarios in it. The result should be an easy-to-use simulation environment in which it is possible to automatically test the Automated Lane-Keeping System provided as a Robot Operating System packages.

ACKNOWLEDGMENT

This work was supported by the Technology Agency of the Czech Republic under the project Certicar CK03000033.

REFERENCES

[1] ECE/TRANS/WP.29/GRVA, *Proposal for the 01 series of amendments to UN Regulation No. 157 (Automated Lane Keeping Systems)*, https://unece.org/sites/default/files/2022-05/ECE-TRANS-WP.29-2022-59r1e.pdf, May 2022.

[2] SAE International, *J3016_202104: Taxonomy and Definitions for Terms Related to Driving Automation Systems for On-Road Motor Vehicles - SAE International*, Apr. 2021. [Online]. Available: https://www.sae.org/standards/content/j3016_202104/ (visited on 08/02/2022).

[3] A. Tenbrock, A. König, T. Keutgens, and H. Weber, "The ConScenD Dataset: Concrete Scenarios from the highD Dataset According to ALKS Regulation UNECE R157 in OpenX," in *2021 IEEE Intelligent Vehicles Symposium Workshops (IV Workshops)*, Jul. 2021, pp. 174–181. DOI: 10.1109/IVWorkshops54471.2021.9669219.

[4] R. Krajewski, J. Bock, L. Kloeker, and L. Eckstein, "The highD Dataset: A Drone Dataset of Naturalistic Vehicle Trajectories on German Highways for Validation of Highly Automated Driving Systems," in *2018 21st International Conference on Intelligent Transportation Systems (ITSC)*, ISSN: 2153-0017, Nov. 2018, pp. 2118–2125. DOI: 10.1109/ITSC.2018.8569552.

[5] A. Dosovitskiy, G. Ros, F. Codevilla, A. Lopez, and V. Koltun, "CARLA: An Open Urban Driving Simulator," in *Proceedings of the 1st Annual Conference on Robot Learning*, ISSN: 2640-3498, PMLR, Oct. 2017, pp. 1–16. [Online]. Available: https://proceedings.mlr.press/v78/dosovitskiy17a.html (visited on 08/04/2022).

[6] "CARLA ScenarioRunner." (2023), [Online]. Available: https://carla-scenariorunner.readthedocs.io/ (visited on 02/17/2023).

[7] S. Riedmaier, T. Ponn, D. Ludwig, B. Schick, and F. Diermeyer, "Survey on Scenario-Based Safety Assessment of Automated Vehicles," *IEEE Access*, vol. 8, pp. 87456–87477, 2020, Conference Name: IEEE Access, ISSN: 2169-3536. DOI: 10.1109/ACCESS.2020.2993730.

[8] P. Weissensteiner, G. Stettinger, J. Rumetshofer, and D. Watzenig, "Virtual Validation of an Automated Lane-Keeping System with an Extended Operational Design Domain," *Electronics*, vol. 11, no. 1, p. 72, Jan. 2022, Number: 1 Publisher: Multidisciplinary Digital Publishing Institute, ISSN: 2079-9292. DOI: 10.3390/electronics11010072. [Online]. Available: https://www.mdpi.com/2079-9292/11/1/72 (visited on 08/04/2022).

[9] "Robot Operating System." (2023), [Online]. Available: https://www.ros.org/ (visited on 02/17/2023).

[10] "CARLA ROS Bridge." (2023), [Online]. Available: https://carla.readthedocs.io/projects/ros-bridge/ (visited on 02/17/2023).

[11] "Nix package manager." (2023), [Online]. Available: https://nixos.org/ (visited on 02/17/2023).

[12] A. Brooks. "Taking off with Nix at FlightAware." (Nov. 2022), [Online]. Available: https://flightaware.engineering/taking-off-with-nix-at-flightaware/ (visited on 12/09/2022).

[13] "Carla simulator Nix packaging." (2023), [Online]. Available: https://github.com/CTU-IIG/carla-simulator.nix (visited on 02/17/2023).

Baseline Selection for Integrated Gradients in Predictive Maintenance of Volvo Trucks' Turbocharger

Nellie Karlsson
School of Information Technology
Halmstad University
Halmstad, Sweden
email: nelkar17@student.hh.se

My Bengtsson
School of Information Technology
Halmstad University
Halmstad, Sweden
email: mypet17@student.hh.se

Mahmoud Rahat
Center for Applied Intelligent Systems Research (CAISR)
Halmstad University
Halmstad, Sweden
email: mahmoud.rahat@hh.se

Peyman Sheikholharam Mashhadi
Center for Applied Intelligent Systems Research (CAISR)
Halmstad University
Halmstad, Sweden
email: peyman.mashhadi@hh.se

***Abstract*—The new advances in Vehicular Systems and Technologies have resulted in a sheer increase in the number of connected vehicles. These connected vehicles use IoT technologies to communicate operational signals with the OEMs, such as the vehicle's speed, torque, temperature, load, RPM, etc. These signals have provided an unprecedented opportunity to adaptively monitor the status of each piece of the vehicle's equipment and discover any possible risk of failure before it happens. This emerging field of study is called predictive maintenance (also known as condition-based maintenance) and has recently received much attention. In this paper, we apply Integrated Gradients (IG), an XAI method until now primarily used on image data, on datasets containing tabular and time-series data in the domain of predictive maintenance of trucks' turbochargers. We evaluate how the results of IG differ, in these new settings, for various types of models. In particular, we investigate how the change of baseline can affect the outcome. Experimental results verify that IG can be applied successfully to both sequenced and non-sequenced data. Contrary to the opinion common in the literature, the gradient baseline does not affect the results of IG significantly, especially on models such as RNN, Long Short Term Memory (LSTM), and GRU, where the data contains time series; the effect is more visible for models like MLP with non-sequenced data. To confirm these findings, and to understand them deeper, we have also applied IG to SVM models, which gave the results that the choice of gradient baseline has a significant impact on the performance of SVM.**

***Index Terms*—Explainable AI (XAI), Predictive Maintenance, Integrated Gradients, Machine Learning.**

I. Introduction

With the increase in popularity of artificial intelligence, several challenges have been brought to light, for example, the lack of transparency, debugging difficulty, lack of control, and biased outcomes that may not represent the real world with its principles and norms [1]. Even though AI is a powerful tool for predictions, it does lack transparency. A significant reason for this is the black-box structure that comes with deep learning methods such as Deep Neural Networks (DNNs), where their hidden layers are hard to visualize for human understanding. In contrast to DNNs and other similarly complex AI models, there are several simpler approaches that are more interpretable and easier to visualize, for example, decision trees; however, they often have limited accuracy [2]. The trade-off between explainability and accuracy can therefore be a challenge.

Because of AI's lack of transparency, it can be challenging to trust the important life-changing decisions the algorithms may take. The AI algorithms and methods give us an answer, but not a *why* or a *how* to that answer. It is hard to trust an algorithm without knowing why and how it made a specific decision. As a result of these challenges, the subject of Explainable Artificial Intelligence (XAI) has arisen. Even though the interest in XAI has increased in the last few years, the term XAI was first coined by Van Lent et al. in 2004 [3]. However, the concept of explainability in machine learning has existed since the 1970s according to [2]. Today, one of the goals of XAI is for humans to understand and trust the reason behind the decisions of an AI model while the model maintains a high prediction accuracy. The theory behind XAI can usually be simplified and divided into four main principles: to justify, to control, to improve, and to discover [2]. This is also a goal for this paper: to explain the reasoning behind the resulting predictions in an understandable way.

There are already several techniques to use for XAI of different kinds, for example, scope-related and/or model-related. Scope-related techniques are divided into two categories: global and local interpretability. Global interpretability is when the technique follows the whole reasoning leading to all of the predictions of the chosen model and understanding its logic. However, global interpretability can be hard to achieve in practice, mainly when it comes to machine learning (ML) models with a large number of parameters. Local interpretability is easier to implement in reality, considering its main focus is on explaining a single prediction and not several. Local interpretability is also the primary approach for the explainability of predictions made by deep neural networks (DNNs) [2].

This paper focuses on explainable AI for Predictive Maintenance (PdM). PdM is a condition-driven preventive method and is used to improve the productivity of a machine by regularly monitoring the parts of the machine to avoid a run-to-failure approach or to maintain a healthy machine [4]–[6].

We have performed several experiments on a real-world turbocharger system dataset provided by Volvo, which is used to predict the remaining useful life. The data is a time-series dataset containing over 400 sensor values and on average 20 timestamps sampled biweekly [7]. To be able to more accurately evaluate the impact of the gradient baseline of integrated gradients (IG) in predictive maintenance, we resort to simulated data. This is due to the fact for the chosen simulated data the feature importance is known to the research community and accordingly easier to evaluate and justify. The first simulated dataset is the Turbofan Engine Degradation Simulation Data Set (CMAPS), which is run-to-failure data that could be used to predict remaining useful life. The dataset contains time as well as sensors reading, which makes the dataset similar to the Volvo dataset [8]. The Tennessee Eastman Process Simulation dataset (ETEPS-CP) is used for the second simulated dataset. The dataset contains information about chemical plants, where some features are measurements while some are manipulated values. This dataset is used as a comparison since it is known which features are measurements and should have more impact on the predictions. The target for the dataset is set to be a classification problem since it is known that the chemical plant runs normally until a fault is induced[[9], [10]]. The dataset contains 54 sensor values and when transformed into time series it contains approximately 38000 timestamps.

A significant difficulty in implementing Integrated Gradients is determining the gradient baseline, which plays an important part in the results. When using images as inputs, it is most common to use a black or white image as a gradient baseline, but the choice of the gradient baseline is not as clear for tabular data. The gradient baseline for tabular data varies depending on the dataset type, and there is not much research on finding the optimal gradient baseline for these types of datasets. We, therefore, want to find a systematic way of defining the optimal gradient baseline for integrated gradients with tabular data as input. The meaning of the word *baseline* in this report refers to the baseline used in integrated gradients, which is explained in more detail in Section III.

This paper explores the following:

1) How the baseline for integrated gradients can be chosen for tabular data in predictive maintenance.
2) How the gradient baseline affects the outcome of different models.

II. Related Work

There are different types of machine learning algorithms and explainable AI that have been used for predictive maintenance. Some algorithms and work that have been adapted to predictive maintenance are *Bagged trees ensemble.*

In the work done by [11], they have used bagged tree ensemble, decision trees and normalized feature deviations to get the model more interpretative. The result of their work concluded that when using bagged trees ensemble, the decision trees as an explanation got a higher quality but did not generate a complete explanation on all test cases as proposed to normalized feature deviations, which got a lower quality of explanation but generated consistent explanations.

In addition to the work described above, the paper [12] evaluates the previous work as well as added LIME to interpret the result. In the paper, they used *Random Under Sampling* (RUS) and boosted trees ensemble, which successfully and correctly classifies all failures as a comparison where the bagged trees ensemble did not.

Other machine learning algorithms used for predictive maintenance and anomaly detection are *Principal Component Analysis* (PCA), *null-space*, *One-Class Support Vector Machines* (OC-SVM), *Extreme Learning Machine* (ELM), and *2 Dimensional Convolutional-based Neural Network Autoencoder* (2D-CNN-AE), which are compared in [13]. The paper compares the different approaches by using the F1-score, where the best approaches are concluded to be the null space and 2D-CNN-AE. Due to the capability of 2D-CNN-AE to detect even small failures, it is outperforming the other methods.

To handle time-series data in prediction, a combination of Convolutional Neural Network (CNN) and Long Short Term Memory (LSTM) is proposed in paper [14]. The paper predicts the remaining lifetime of aircraft engines by comparing Convolutional Neural Network (CNN), LSTM, and a combination of the two algorithms, CLSTM, where the result is that the combination of the two algorithms provides the highest accuracy rate. Another work that handles both CNN and LSTM together is a fault diagnostic system on mechanical data from a gearbox [15]. The result of the algorithm proposed is 97% accuracy, where the algorithm is able to detect which fault is detected.

III. Integrated Gradients

Integrated Gradients (IG) is a technique for model interpretability used to visualize the relationship between the prediction of the model and the input features, often when the input is an image. Similar to SHAP, IG is also inspired by game theory, especially the Aumann-Shapley value, which SHAP is based on [16]. Below, the way to compute IG is shown as well as in eq. 1:

1) The first step is to identify the input and output. In this study, the input is the sequential data whereas the last layer of the model is the output.
2) To be able to identify features that are important to the prediction of the neural network, the second step is to choose a gradient baseline as an input.

3) The third step is then to interpolate the chosen gradient baseline for a number of steps. The number of steps is a hyperparameter and represents the number of steps needed for the given input in the gradient approximation, where the recommended number is between 20-1000 steps,
4) After the gradient baseline has been interpolated they are preprocessed, and then a forward pass is done.
5) Lastly, the gradients for the interpolated data points are obtained and then by using the trapezoidal rule, the gradients integral is approximated.

$$IG_i^{approx}(x) :== (x_i - x_i^{'}) \times \sum_{k=1}^{m} \frac{\partial F(x^{'} + \frac{k}{m} \times (x - x^{'}))}{\partial x_i} \times \frac{1}{m} \quad (1)$$

Where:

$[x_i]$ = Input Data
$[x_i^{'}]$ = Gradient Baseline
$[m]$ = Number of Steps in the Integral Approximation

There are several advantages of integral gradients. An example is sensitivity, which means that it will give a non-zero attribution every time there is a difference in one feature between input and gradient baseline but also a difference in predictions. Another example is the invariance of implementation, where two models' feature attributions will be the same if they both are functionally equivalent, without regard to the network architecture [17].

IV. METHODOLOGY

A. Setup

Before the experiments are performed, the data needs to be prepared, which is done differently among the data sets. For the CMAPS dataset, standard scaling and hyperparameter tuning are used. For the ETEPS-CP dataset, standard scaling was used to preprocess the data. For the Volvo dataset, more preparations need to be taken, where the first step is to handle the NAN values by imputation with the mean values of each column. It is important to note that Mean Imputation (MI) can lead to biased estimates and predictions, especially if the number of NAN values is significant. Before the imputation, the dataset only contained approximately 2% of NAN values, and therefore we decided that MI is a suitable type of imputation in this dataset.

After MI, the columns containing objects or strings are label-encoded so that the model used only has float or integer as input values. Most of the ML models do not take strings as inputs, which is why label encoding is needed.

Lastly, the data needs to be scaled and normalized when employing deep learning models. For this, MinMaxScaler and standardScaler from the scikit-learn library are used.

B. Machine Learning Methods

To be able to evaluate the XAI methods, as well as the gradient baselines, we first need to implement reliable models. The models used in the experiments are Recurrent Neural Network (RNN), Long short-term memory (LSTM), Gated Recurrent Unit (GRU), and Multilayer Perceptrons (MLP). Support Vector-Machine (SVM) is also implemented to affirm the results we receive from the experiments on MLP, which can be seen in section V. These regressor models are evaluated using Mean Absolute Error (MAE) and Root Mean Squared Error (RMSE). For the classification models, the measurements will be on accuracy as well as Area Under the Receiver Operating Characteristic Curve (ROC AUC) where it should present better than randomly choosing a class, which is better than 0.5. The models perform better than the baseline, which means that for the regression part, the mean absolute error is smaller than the mean value for the test sets. For the classification, the accuracy is above 50%, and the ROC AUC is over 0.5.

C. Systematic Choice of Baseline

The choice of a gradient baseline has a large impact on the result of Integrated Gradients, and it is therefore crucial to have an appropriate gradient baseline. In figure 1 we provide a systematic approach to the choice of gradient baseline, which is explained in more detail in the following paragraphs.

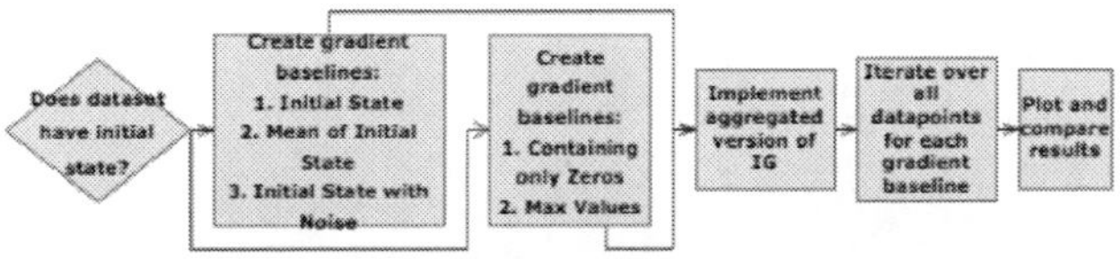

Fig. 1: A systematic approach on how to select gradient baseline.

For the Volvo dataset and the simulated datasets, the gradient baselines used are a gradient baseline with the Initial State (IS), the Mean Initial State (MIS), and the Initial State with Gaussian Noise (ISN). For datasets without an initial state, gradient baselines consisting of either only zeros or max values can be used to resemble an image. Our approach of using a gradient baseline with only zeros simulates a white image, and a gradient baseline with the max value for all features simulates a black image.

An aggregated version of integrated gradients has been used to get an overall view of the attributions. The aggregated model of integrated gradients is an iterative type of integrated gradients, where all attributions for all data points are iterated. The reason for using an aggregated integrated gradient is to get an average of all feature importance.

For all datasets, the aggregated version of integrated gradients was iterated over all data points for the different gradient baselines. The different baselines used are the initial state of the data, the initial state with gaussian noise as well as the mean value of all the states. The initial state is used as a ground truth of the datasets, where the different variations of the ground truth are used to explore the result when using different gradient baselines.

Lastly, the results between the different gradient baselines are plotted and compared, where the gradient baseline with the most logical results (in accordance with domain experts) is the most favorable choice of baseline.

V. RESULTS

A. Gradient Baseline

The main experiment is to find the optimal gradient baselines for sequential and tabular datasets. We use the two simulated benchmark datasets (ETEPS-CP and CMAPS) to test gradient baselines similar to the ones used on the Volvo dataset.

1) Simulated Data - CMAPS: Figure 2 shows the features that appeared as the top three features with the most importance according to IG for all of the RNN models. In Figure 2, we can see that sensor 11 plays a significant role in the prediction of all models, as it almost always is the third most important feature. We can also see that sensor 4 is clearly an important feature for all of the models, as it rather often placed as the second most important feature according to IG.

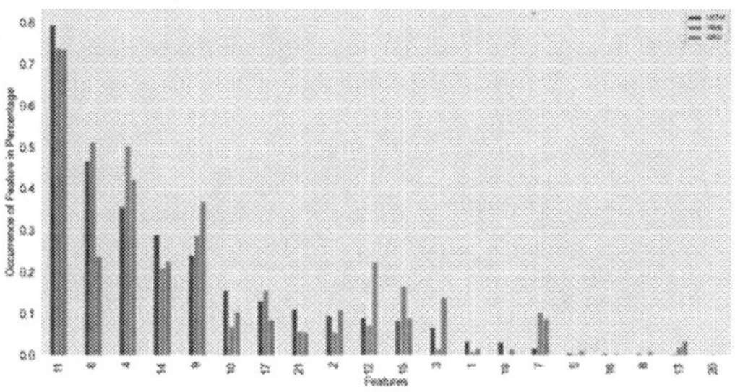

Fig. 2: Feature occurrence for top three features with the most importance in percentage according to the results of IG. These results are for the models LSTM, RNN, GRU on the CMAPS dataset, and depend on the model.

Overall, all RNN models (RNN, LSTM, and GRU) seem to give similar results when using integrated gradients, where sensors 11, 4, 6, and 9 have the most significant impact on the prediction. Looking at Figure 3 with the MLP model, we also receive sensors 11 and 4 at the top. Sensors 6 and 9, however, can only be seen towards the far right of the graph, occurring under 5% as the top 3 features.

Looking at Figure 3, the results between the gradient baselines of the MLP model also vary more than between the RNN models. Here, we can see that the gradient baselines pay a different amount of attention to a larger variety of sensors than the RNN models do in Figure 2. We can also see this in Figures 4, 5 and 6, where how large of an impact that the different gradient baselines have on each model.

In the case of the CMAP dataset, it could be that the MLP model is sensitive toward specific sensors, such as sensors 6 and 9 (as explained above). This sensitivity could lead to the baselines playing a much more significant role in the results of IG.

Another reason the results between the baselines differ much more for the MLP model than for the RNN models is that the baseline may not have as much significance when using models specified toward sequenced data. We, therefore, theorize that when using a model such as RNN where the data is sequenced, the baseline does not affect the outcome of IG to a large extent, as long as the baseline is a reasonable one (for example, the initial state). For models where the input is not sequenced, the gradient baseline affects the results of IG more.

To endorse this theory, we applied IG to a Support Vector Machine (SVM) model where the input is not sequenced. In Figure 7, we can see that the baselines play a significant role in the outcome of IG, similar to the results of MLP.

2) Simulated Data - ETEPS-CP: The results for the MLP model can be seen in Figure 10. The initial state is almost identical to the initial state with noise. However, the mean of the initial states gives completely different results. The similarities between the initial state and the initial state with noise could be that adding noise to the baseline does not change the baseline significantly or that these features are highly correlated to the prediction. When comparing to Figure 10, it is seen that the initial state and initial state with noise is similar, which also was seen in the table. Moreover, the mean initial state pays attention to a broader number of features, with a smaller number of features occurring more than others. The conclusion to draw from the dataset with an MLP model is that the initial state and initial state with noise perform better since fewer features occur in the top for around 50% of all data points.

In Figure 10, it is shown that the initial state and initial state with noise pay attention to the same features while the mean initial state pays attention to multiple features.As seen in Figure 10, the gradient baseline MIS is not shown in the figure. This is because the values from MIS are giving feature values near zero for the MLP model. When looking at the other gradient baselines, it is clear that there is no difference between the values for the feature importance.

The results for the GRU model can be seen in Figure 13. All three baselines are similar to each other in both placement and occurrence.In Figure 13, the values for the top features are similar for every gradient baseline. Compared to the MLP-model, Figure 10, which has more features occurring at the top, the GRU model gives fewer features with no difference between baselines.

The results for the LSTM model can be seen in Figure 12. However, the different features do not appear in the same placement or occurrence. In Figure 12, the values for the top features is similar for every gradient baseline as for the GRU model, Figure 13.

The results for the RNN model can be seen in Figure 11. As seen in Figure 11, the different baselines are almost identical to each other. Further looking into both GRU in Figure 13 and LSTM in Figure 12, the same pattern reoccurs, where all the different baselines give almost identical results for the same model and baselines.

However, comparing the different time-series models seen in Figure 8, the models pay attention to the same features but

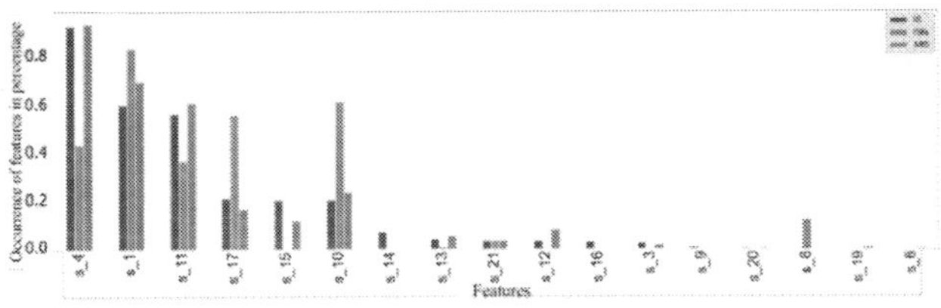

Fig. 3: Feature occurrences for top three features with the highest importance (in percentage), according to the results of IG, for the four different types of networks. These results are for the CMAP dataset for MLP model, and depend on the baselines.

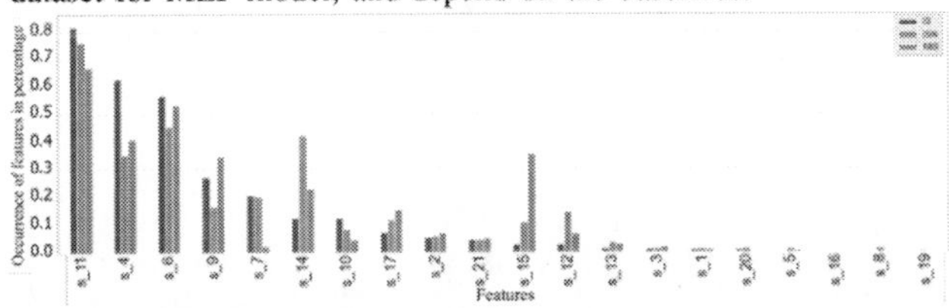

Fig. 4: Feature occurrences for top three features with the highest importance (in percentage), according to the results of IG, for the four different types of networks. These results are for the CMAP dataset for RNN model, and depend on the baselines.

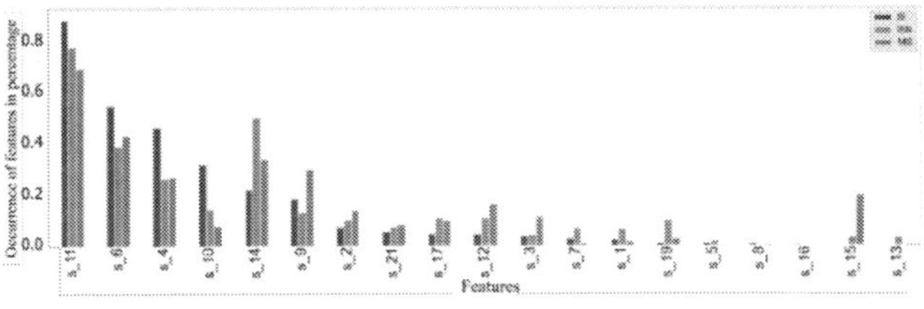

Fig. 5: Feature occurrences for top three features with the highest importance (in percentage), according to the results of IG, for the four different types of networks. These results are for the CMAP dataset for LSTM model, and depend on the baselines.

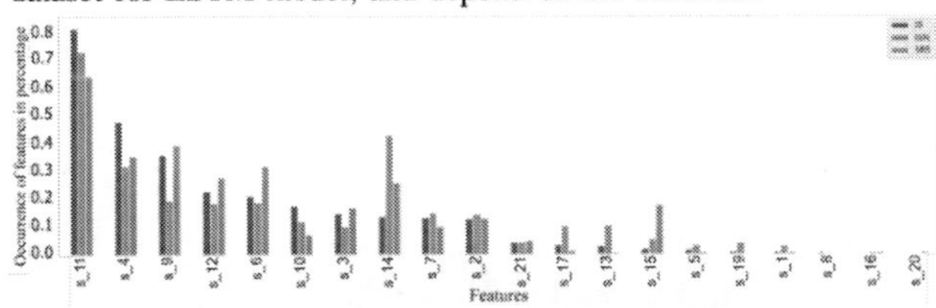

Fig. 6: Feature occurrences for top three features with the highest importance (in percentage), according to the results of IG, for the four different types of networks. These results are for the CMAP dataset for GRU model, and depend on the baselines.

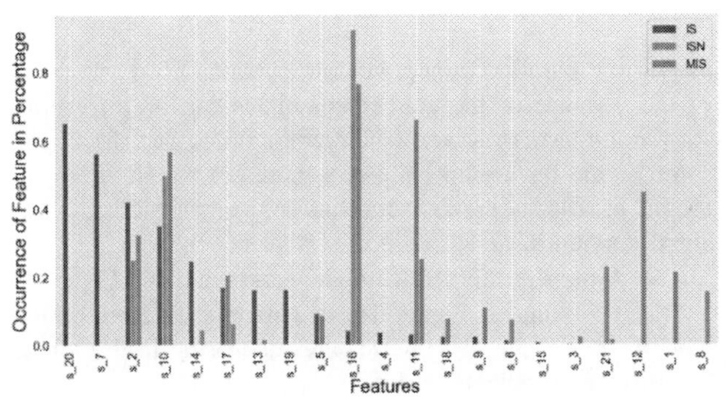

Fig. 7: Feature occurrence for top three features with most importance in percentage according to the results of IG. These results are for the model SVM on the CMAP dataset, and depend on the models.

with different magnitudes.

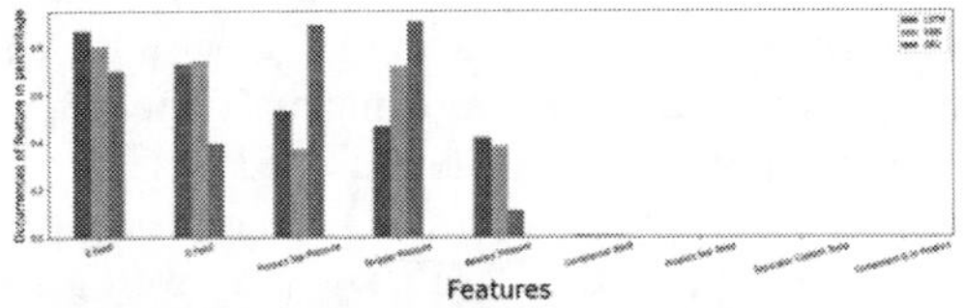

Fig. 8: Feature occurrence for top three features with most importance in percentage for IG. These results are for the models RNN, GRU and LSTM on the ETEPS-CP dataset.

The result for IG on the sequential data is once again tested as in V-A1, where similar results are presented regarding sequential data. To see if the theory that the gradient baseline affects the results of IG on non-sequential models more than sequential models also applies to the ETEPS-CP dataset, we apply SVM classification. The result of the SVM Classification can be seen in Figure 12 which strengthens the belief that the non-sequential model is more affected by the gradient baseline.

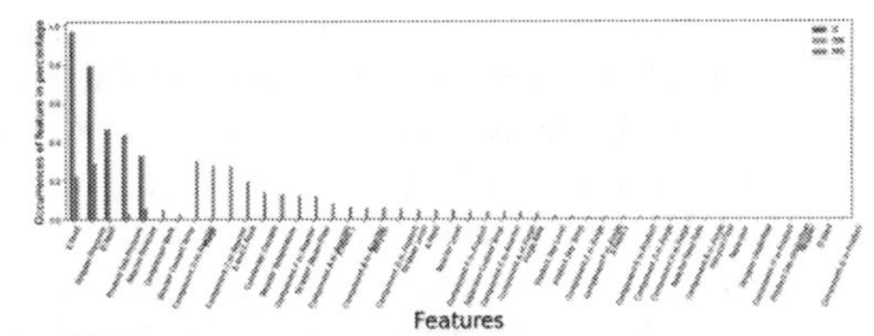

Fig. 9: Feature occurrence for top three features with the most importance in percentage according to the results of IG. These results are for the SVM Classification model on the ETEPS-CP dataset, with three different baselines.

In Figure 8, the feature occurrence from IG for LSTM, GRU, and RNN models can be seen. From Figure 8, it is shown how the different models give similar results to the IG.

The overall conclusion for the ETEPS-CP dataset is that the choice of baselines does not seem to affect the explanation for the time-series models. Moreover, when comparing the MLP to time-series, the gradient baseline significantly impacts the explanations.

3) Volvo Dataset: The results for the MLP model can be seen in Figure 14.

In Figure 14 the occurrence of the features can be seen. As with the simulated datasets, MLP pays attention to multiple features depending on the gradient baseline. Looking into the values in Figure 14, it can be seen that the initial state and initial state with noise are similar. However, the mean initial state pays attention to a broader number of features.

Evaluating Figure 14, shows that they both often occur and have a large impact on the prediction. For the MLP model,

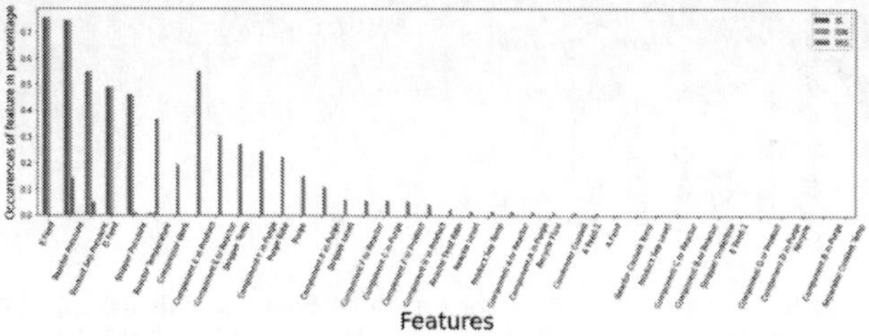

Fig. 10: Feature occurrence for top three features with the most importance in percentage according to the results of IG. These results are for the ETEPS-CP dataset for MLP model, and depend on the baselines.

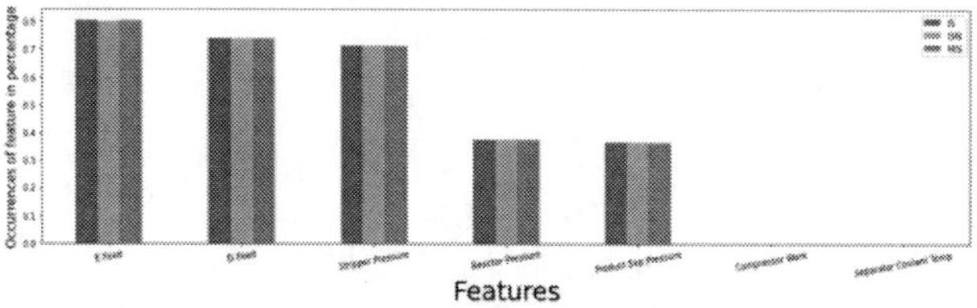

Fig. 11: Feature occurrence for top three features with the most importance in percentage according to the results of IG. These results are for the ETEPS-CP dataset for RNN model, and depend on the baselines.

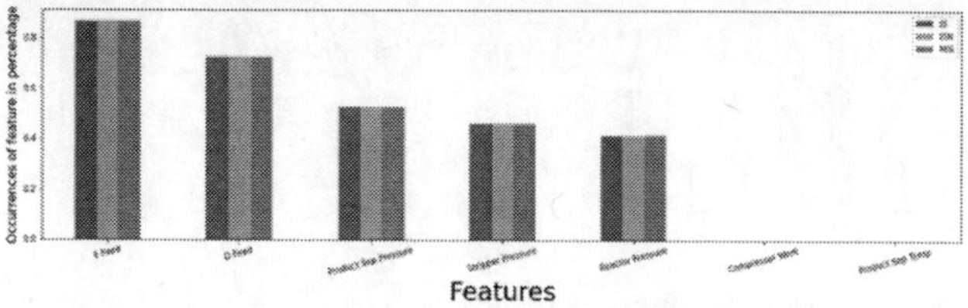

Fig. 12: Feature occurrence for top three features with the most importance in percentage according to the results of IG. These results are for the ETEPS-CP dataset for LSTM model, and depend on the baselines.

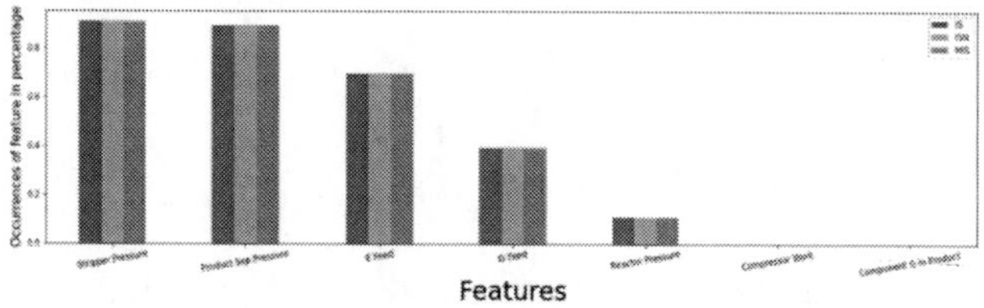

Fig. 13: Feature occurrence for top three features with the most importance in percentage according to the results of IG. These results are for the ETEPS-CP dataset for GRU model, and depend on the baselines.

the initial state is similar to the initial state with noise, while the mean initial state has a broader view of features.

The MLP-model, as seen in Figure 14, shows a broad spectrum of features where only the mean initial state has features that occur in the top three at more than 50% of the dataset. Some similarities can be seen between the initial state and the initial state with noise. However, the other datasets are not identical, which could result from having a more complex dataset. IG is also very sensitive to noise, leading to the difference between the IG for different datasets.

When comparing the baselines for the RNN model, as seen in Figure 15, the different baselines do not differ as much as for the MLP. However, the different baselines do not seem to have a huge impact on important features. By looking into GRU, Figure 17 as well as LSTM, Figure 16 the same patterns are occurring. For all time-series data there are not any significant features that occur in the top three.

The overall conclusions that can be drawn from this are how sensitive IG is towards the noise and that the baselines do not significantly impact time-series data, which is shown in sections V-A1 and V-A2. Some features occur in multiple gradient baselines; however, no feature occurs in all three different baselines for the MLP-model.

The results for the RNN model can be seen in Figure 15. IG gradient baseline initial state and initial state with noise are similar. In contrast, the mean initial state pays attention to more features, which means that the theory of gradient baseline can be applied when the dataset is robust and does not have noise since IG is sensitive to noise.

The results for the GRU model can be seen in Figure 17.

Evaluating Figure 17, it can be seen that for all gradient baselines the model pays attention to multiple features depending on the baseline. Since the data contains noise, the result of the variance of the gradient baselines can be disregarded for the outliers. However, when looking closer at the figures, some of the features occur in all baselines which strengthen the belief that the gradient baseline for sequenced data with a small amount of noise result in similar conclusions.

The results for the LSTM model can be seen in Figure 16.

When looking at Figure 16, it can be seen that the data contains noise due to some features appearing in the top ten for only one gradient baseline.

VI. Conclusions & Future Work

In this paper, we have dived deeper into the XAI method integrated gradients to justify the predictive maintenance results for a dataset provided by Volvo. To our knowledge, IG is a method commonly used for data of images and not the time-series data that the Volvo dataset contains. The lack of work done on time-series data for IG can be because the choice of gradient baseline can be seen as complex. Therefore, we have focused the paper on how to find a good baseline and how the baseline affects the result depending on the deep learning model and the type of data. We have also investigated other types of XAI methods to either justify or compare the results of our experiments.

We observed that integrated gradients are a good method to interpret the behavior of deep learning models in predictive maintenance, especially for time series data. However, we still believe that the choice of gradient baseline continuous to be seen as difficult. As stated before, we theorize that the gradient baseline's effect on the results of integrated gradients decreases for RNN models with sequential data and increases on models like MLP with non-sequential data. However,

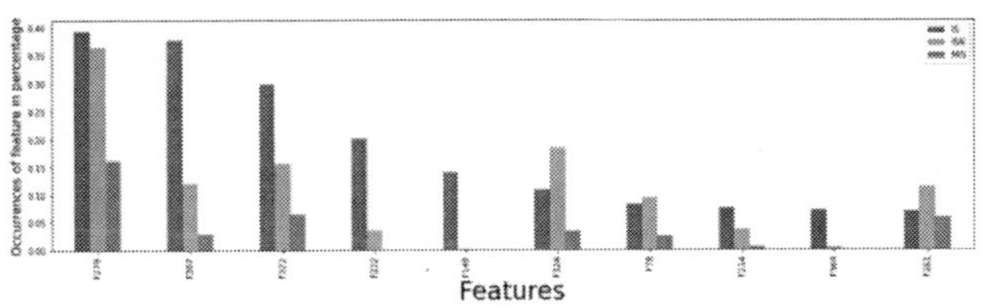

Fig. 14: Feature occurrence for top three features with the most importance in percentage according to the results of IG. These results are for the Volvo dataset for MLP model, and depend on the baselines.

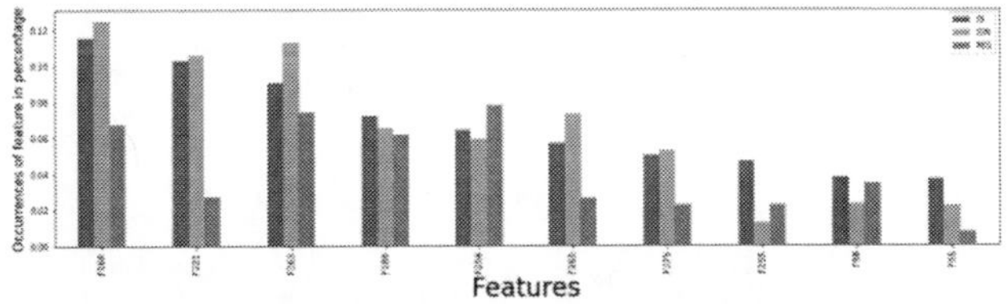

Fig. 15: Feature occurrence for top three features with the most importance in percentage according to the results of IG. These results are for the Volvo dataset for RNN model, and depend on the baselines.

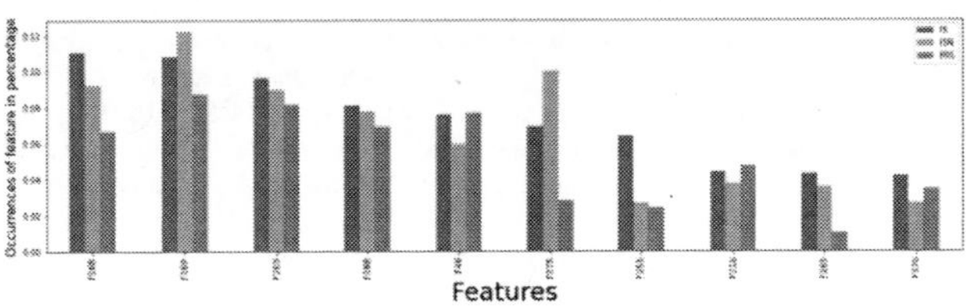

Fig. 16: Feature occurrence for top three features with the most importance in percentage according to the results of IG. These results are for the Volvo dataset for LSTM model, and depend on the baselines.

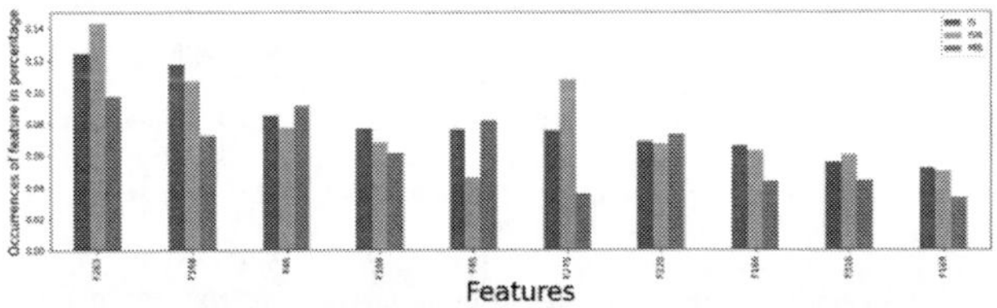

Fig. 17: Feature occurrence for top three features with the most importance in percentage according to the results of IG. These results are for the Volvo dataset for GRU model, and depend on the baselines.

integrated gradients are sensitive to noise, and therefore, the results can vary depending on the dataset. Even a dataset with noise can get a similar result as the theory depending on the model. To strengthen this theory, we applied integrated gradients with three different gradient baselines on SVR and SVM models with non-sequential data, which gave similar results as it did for the MLP model.

To answer the questions 1. "*How can the baseline for integrated gradients be chosen for tabular data in predictive maintenance?*" and 2. "*How does the baseline affect the outcome of different models?*", we have applied IG on several datasets and models, with different gradient baselines. However, since IG is a local XAI method, we implemented an aggregated version of the method to get a larger view of how the model makes its predictions. This makes it easier to see how the features impact the whole model, and not only one prediction, which is very useful in predictive maintenance. We can see from the experiments and conclusions that the gradient baseline has a more significant impact on the results of IG on tabular and non-sequential data on models such as MLP and SVM. In contrast, the impact of the gradient baselines decreases for sequential models such as RNN, LSTM, and GRU. These results could imply that the IG method is more suited for sequential data. Furthermore, we observe that applying IG to a deep learning model provides knowledge on the importance of the features differently depending on how robust the data is when using data with time series. With noisy data, such as the Volvo dataset, the IG has a more challenging time making clear conclusions. This is not anything new, considering we know that IG is sensitive to noise; however, it is now more evident that this also applies to data with time series and not only for images.

We would have liked to discuss the results with domain experts, which is something we will bring for future work. We would also like to see a combination of integrated gradients with another gradient-based method as a future work within the area of XAI and predictive maintenance for time-series data.

Acknowledgment

This work was supported by research grants from KK-Foundation, Sweden.

References

[1] B. Shukla, I.-S. Fan, and I. Jennions, "Opportunities for explainable artificial intelligence in aerospace predictive maintenance," in *PHM Society European Conference*, vol. 5, pp. 11–11, 2020.

[2] A. Adadi and M. Berrada, "Peeking inside the black-box: A survey on explainable artificial intelligence (xai)," *IEEE Access*, vol. 6, pp. 52138–52160, 2018.

[3] M. Van Lent, W. Fisher, and M. Mancuso, "An explainable artificial intelligence system for small-unit tactical behavior," in *Proceedings of the national conference on artificial intelligence*, pp. 900–907, Menlo Park, CA; Cambridge, MA; London; AAAI Press; MIT Press; 1999, 2004.

[4] M. Rahat, S. Pashami, S. Nowaczyk, and Z. Kharazian, "Modeling turbocharger failures using markov process for predictive maintenance," in *30th European Safety and Reliability Conference (ESREL2020) & 15th Probabilistic Safety Assessment and Management Conference (PSAM15), Venice, Italy, 1-5 November, 2020*, European Safety and Reliability Association, 2020.

[5] V. Revanur, A. Ayibiowu, M. Rahat, and R. Khoshkangini, "Embeddings based parallel stacked autoencoder approach for dimensionality reduction and predictive maintenance of vehicles," in *IoT Streams for Data-Driven Predictive Maintenance and IoT, Edge, and Mobile for Embedded Machine Learning*, pp. 127–141, Springer, 2020.

[6] M. G. Altarabichi and et al., "Stacking ensembles of heterogenous classifiers for fault detection in evolving environments," in *30th European Safety and Reliability Conference, ESREL 2020 and 15th Probabilistic Safety Assessment and Management Conference, PSAM15 2020, Venice, Italy, 1-5 November, 2020*, pp. 1068–1068, Research Publishing, 2020.

[7] M. Rahat, P. S. Mashhadi, S. Nowaczyk, T. Rognvaldsson, A. Taheri, and A. Abbasi, "Domain adaptation in predicting turbocharger failures using vehicle's sensor measurements," in *PHM Society European Conference*, vol. 7, pp. 432–439, 2022.

[8] A. Saxena and K. Goebel, "Turbofan engine degradation simulation data set," *NASA Ames Prognostics Data Repository*, pp. 1551–3203, 2008.

[9] C. Reinartz, M. Kulahci, and O. Ravn, "An extended tennessee eastman simulation dataset for fault-detection and decision support systems," *Computers & Chemical Engineering*, vol. 149, p. 107281, 2021.

[10] C. C. Reinartz, M. Kulahci, and O. Ravn, "Tennessee Eastman Reference Data for Fault-Detection and Decision Support Systems," 2021.

[11] S. Matzka, "Explainable artificial intelligence for predictive maintenance applications," in *2020 Third International Conference on Artificial Intelligence for Industries (AI4I)*, pp. 69–74, 2020.

[12] A. Torcianti and S. Matzka, "Explainable artificial intelligence for predictive maintenance applications using a local surrogate model," in *2021 4th International Conference on Artificial Intelligence for Industries (AI4I)*, pp. 86–88, 2021.

[13] O. Serradilla, E. Zugasti, J. Ramirez de Okariz, J. Rodriguez, and U. Zurutuza, "Adaptable and explainable predictive maintenance: Semi-supervised deep learning for anomaly detection and diagnosis in press machine data," *Applied Sciences*, vol. 11, no. 16, 2021.

[14] A. P. Hermawan, D.-S. Kim, and J.-M. Lee, "Predictive maintenance of aircraft engine using deep learning technique," in *2020 International Conference on Information and Communication Technology Convergence (ICTC)*, pp. 1296–1298, 2020.

[15] T. Haj Mohamad, A. Abbasi, E. Kim, and C. Nataraj, "Application of deep cnn-lstm network to gear fault diagnostics," in *2021 IEEE International Conference on Prognostics and Health Management (ICPHM)*, pp. 1–6, 2021.

[16] P. Sturmfels, S. Lundberg, and S.-I. Lee, "Visualizing the impact of feature attribution baselines," *Distill*, 2020. https://distill.pub/2020/attribution-baselines.

[17] Y. B. M.R. Gazarra, D.Singh, "Enhanced integrated gradients: improving interpretability of deep learning models using splicing codes as a case study," *Genome Biology*, vol. 21, no. 149, 2020.

An Efficient YOLOv7x Based Automated Street Parking Space Detection for Smart Cities

Tala Bazzaza, Hamid Reza Tohidypour, Yixiao Wang, and Panos Nasiopoulos
Electrical & Computer Engineering, Univeristy of British Columbia
Vancouver, BC, Canada
email: tbazzaza@student.ubc.ca, {htohidyp, yixiaow, panosn}@ece.ubc.ca

Abstract— **Finding available street parking spots is a cause of increased traffic in metropolitan cities. To address this challenge, in this paper, we propose a unique real-time street parking detection scheme that utilizes visual information and object recognition to accurately detect empty street parking spots. We also introduce a comprehensive video dataset that is captured specifically for this task and is used for training our networks. Among several network options for localization, our tests on YOLOv7 achieved the highest accuracy and speed, making it an ideal choice for real-time street parking detection for human driven as well as autonomous vehicles.**

Keywords: street parking detection; deep learning, YOLOv7; real-time performance; object recognition.

I. INTRODUCTION

As cities continue to grow and urbanize, traffic congestion has become an increasingly common problem. In metropolitan cities, it is estimated that 30-50% of traffic congestion is caused by drivers searching for parking spots during peak hours [1][2]. This not only leads to waste of fuel and increased levels of pollution, but also to a significant reduction in productivity due to driving aimlessly multiple times around city blocks in search of a vacant parking spot [3]. To address this issue and to improve traffic management, an efficient and functional street parking detection system is needed to deploy the vision of smart cities. This system will direct drivers towards vacant parking spots around the block, therefore reducing unnecessary delays which worsen traffic conditions [4][5]. Previously, methods that rely on ultrasonic sensors were used to quantify the target area by using a virtual grid map and establishing a coordinate system for parking spot detection [6]. Other methods have utilized autonomous sensor nodes with Wireless Sensor Networks (WSNs) for monitoring parking occupancy in lots [7][8]. A previous study utilizes video surveillance camera data to detect parking spots using Support Vector Machines (SVM) and k-nearest neighbors algorithms [9]. Although this method produced results with high accuracy, it is not practical for on-street parking detection as it uses aerial views captured by video surveillance cameras which do not necessarily cover the city streets. Nowadays, many researchers are using computer vision techniques and deep learning methods to detect available parking spots. In [10], the authors use instance segmentation algorithms and convolutional neural networks to perform real time processing on data to determine if the parking spot is vacant. These methods are also more scalable and robust than the traditional sensor-based systems, as they can work under different weather conditions, lighting conditions, and camera angles. However, there are still some challenges that need to be addressed, such as occlusions, different parking spot sizes and shapes, and variations in parking spot markings.

This paper explores the use of the latest You Only Look Once (YOLO) network architecture, namely YOLOv7, for accurately detecting available street parking spaces. To achieve this, we have created a new and extensive video dataset of city street parking spaces and we have trained and fine-tuned the network on this dataset. We compare the performance of our network against the state-of-the-art approach presented in [11], which is based on YOLOv4. Evaluation results show that our YOLOv7 outperforms YOLOv4 in terms of Mean Average Precision (mAP) at different threshold levels of overlap between the predicted bounding box and the ground-truth bounding box. More specifically, YOLOv7 reached a mAP of 89.9% at a threshold of 50% overlap, while YOLOv4 reached a mAP of 83.3% at the same threshold. Additionally, YOLOv7 showed faster inference time and better generalization ability than YOLOv4. These results demonstrate the superiority of YOLOv7 and pave the way for its use in real-time street parking detection for human driven as well as autonomous vehicles.

The rest of this paper is organized as follows. Section II describes the methodology of our proposed method. Section III describes the results and evaluation. In Section IV, conclusion and future work are presented. The acknowledgement closes the article.

II. OUR PROPOSED METHOD

A. *Dataset and labelling*

In this study, we used a dataset consisting of 55 videos captured by our team in the city of Vancouver, Canada. The dataset was carefully curated to include a diverse range of weather conditions (sunny, cloudy, rainy, snowy) and location scenarios. To prepare the dataset for training and evaluation, we utilized the Fast Forward Moving Picture Experts Group (ffmpeg) tool to extract frames from the videos, which were then labeled using the Computer Vision Annotation Tool (CVAT) software [12][13]. This allowed us to create a dataset that is representative of real-world scenarios and provides a robust evaluation of the performance of the object detection models.

In our labeling technique, we decided to implement a single class, labeling only available parking spaces. We labeled parking spots that are within a distance of 5 meters from the car, and only focused on the right side of the street. This approach helps eliminate double counting parking spots and more accurately determining if the parking space is long enough to fit a car. In addition, using unlabeled data during training can introduce the model to learn features that are not related to vacant parking spots such as intersections, bus stops, yellow curbs, and non-vacant parking spots. These unlabeled frames

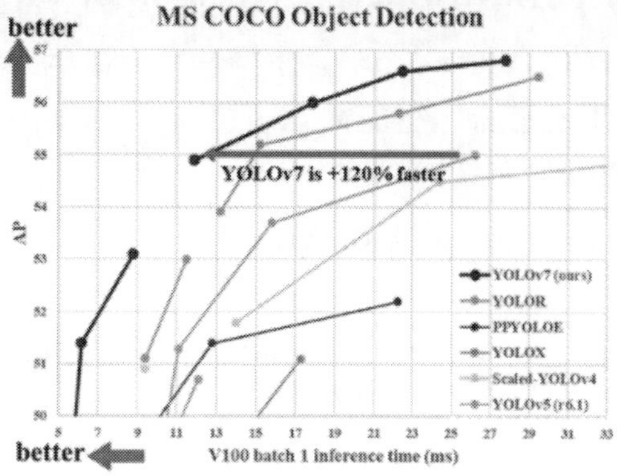

Figure 1. Comparison of YOLOv7 with previous object detection networks [16].

were included along with empty text files as there are no coordinates highlighted. We created a dataset that contains a total of 3381 frames of city street parking spaces. It is well established that a balanced dataset of labeled and empty (no-label present) frames yields the best training results for object detection [14][15]. For this reason, the dataset was composed of 1776 frames that were labeled as available parking spots and 1605 frames that were left unlabeled (empty text files) to be used during the training process. This dataset was split in 85% for training and 15% for validation. Additionally, a different and unseen dataset consisting of 278 frames from the city of Vancouver was used for testing. The testing dataset is used to evaluate the performance of our proposed model.

B. Our network

We chose the YOLOv7 family architecture as the basis for our network. It is worth noting that YOLOv7 has made significant advancements to the previous YOLO models both on the architectural level and at the trainable bag-of-freebies level, which improves the accuracy of the model without increasing the cost of training. Overall, YOLOv7 has a more efficient architecture that reduces the number of parameters by 75%, thus requiring 36% less computation and achieving 1.5% higher Average Precision (AP) compared to the previous models [15]. Figure 1 compares YOLOv7 with other real-time object detectors. We observe that YOLOv7 achieves state-of-the-art performance with improved accuracy and lower complexity.

YOLOv7, like the entire family of this architecture, uses data augmentation techniques to increase the size of the training dataset and improve the generalization of the model, which as a result avoids possible overfitting. However, since in our implementation we are mainly focused on the right side of the street when locating parking spots, some augmentation techniques like flipping and rotation are not applicable as they will produce erroneous images (i.e., parking on the left side or cars upside down) and for this reason they were disabled. Nonetheless, we implemented a modified version of the Mosaic data augmentation introduced by the inventors of YOLOv4, selectively combining four images to generate a new one, but making sure that we are not violating our requirement to have parking spots only on the right side [14]. This technique proved to help our YOLOv7 based network to learn more features and become robust to different lighting conditions, camera angles, and object scales.

We decided to train two versions of the YOLOv7 family, the original YOLOv7 and the latest YOLOv7x. The main differences between the two is that YOLOv7 uses the method of stack scaling on the neck, which is a technique to increase the capacity of a model by adding more layers to it. In this case, this technique is applied to the "neck" of the model, which is a key component of the architecture that helps to extract features from the input image. By stacking more layers on the neck, the model is made more powerful and able to detect more complex objects.

On the other hand, YOLOv7x in addition to the neck scaling scheme of YOLOv7 performs compound scaling on the neck, which increases the depth and width of the entire model simultaneously, as opposed to only increasing one or the other, leading to an improved performance.

Anchor boxes are a key component of object detection algorithms, as they are used to predict the location and size of objects in an image. YOLOv7 uses an auto-anchor algorithm borrowed from YOLOv5, which adapts to the scale of the objects in an image by using a single anchor box that can adjust to different scales [15][17]. To this end, before the training process begins, the suitability of the provided anchors for the dataset is evaluated. If the fit is not optimal, new anchors are recalculated that are more suitable for the data. The model is then trained using these newly generated, more appropriate anchors.

III. Results and Evaluation

We compared our suggested network with the state-of-the-art method presented in [11]. In order to fairly evaluate the performance of the YOLOv4 network used in [11], we had to retrain it using our new and more comprehensive dataset.

We had to address the limitation of YOLOv4 in generating anchor boxes by using k-means clustering. Since we are using a custom dataset, we generated anchor boxes based on the aspect ratio and scale of the objects in our dataset before starting the training process in YOLOv4. The new calculated anchor boxes were added manually in each of the yolo-layers while configuring our model.

Regarding our proposed approach, we first trained our YOLOv7 and YOLOv7x networks using the computing clusters available by Compute Canada [18]. We started training YOLOv4, YOLOv7 and YOLOv7x with the pretrained weights of the darknet framework and the pretrained weights of YOLOv7 and YOLOv7x [14][15].

Performance evaluation and accuracy of three models is done using the mean Average Precision (mAP) metric. Average precision is calculated by measuring the precision and recall of the model at different intersection-over-union (IoU) thresholds,

Table I. Validation Results of all the models

Model	mAP @ 0.5
YOLOv4	82%
YOLOv7	84%
YOLOv7x	90%

TABLE II. TESTING RESULTS OF ALL THE MODELS

Model	Precision	Recall	mAP @ 0.5
YOLOv4	0.84	0.81	83.3%
YOLOv7	0.90	0.79	86.5%
YOLOv7x	0.91	0.81	89.9%

which is the ratio of the area of overlap between the predicted bounding box and the ground-truth bounding box to the area of the union of the two boxes. In this paper, we compare the models at an IoU threshold of 0.5. Table I shows the performance of our trained models on the validation set.

We observe that the YOLOv4 model scored a mean Average Precision of 82% for the validation set. On the other hand, both versions of YOLOv7 outperformed YOLOv4, achieving mAP@0.5 of 84% and 90%, respectively.

We also observed that YOLOv7x achieved the best weights at (mean Average Precision) mAP@0.5 with levels of 90% - see Fig. 2 that shows the precision-recall curve.

In order to evaluate the performance of our models against that of YOLOv4 for unseen test data, we tested all of them on 278 previously unseen frames captured by our lab in the city of Vancouver. Table II shows results of detecting vacant parking spots. We observe that YOLOv7x achieves the best performance, reaching a mAP of 89% at a threshold of 50% overlap between the labeling bounding box and the predicted one. Overall, our models performed significantly well in all different areas of the road such as main street, crosswalks, and intersections. and other type of side entrances that could be confusing even for a human. Figure 2 below shows the precision-recall curve for the testing set of YOLOv7x.

Figs. 3a and 3b show two representative examples of street parking detection performed by YOLOv4 and YOLOv7x respectively. It is obvious from Fig.3a that the YOLOv4 model was unable to detect parking spots in some instances where the car was driving on the same lane as the parking lane (no bounding box is present). However, Fig. 2b shows that the YOLOv7x model accurately detected the available parking spaces (purple bounding box). The Intersection of Union (IoU) score displayed above the bounding boxes represents the degree of overlap between the labeled bounding box and the one predicted by the model. As our goal is to detect street parking

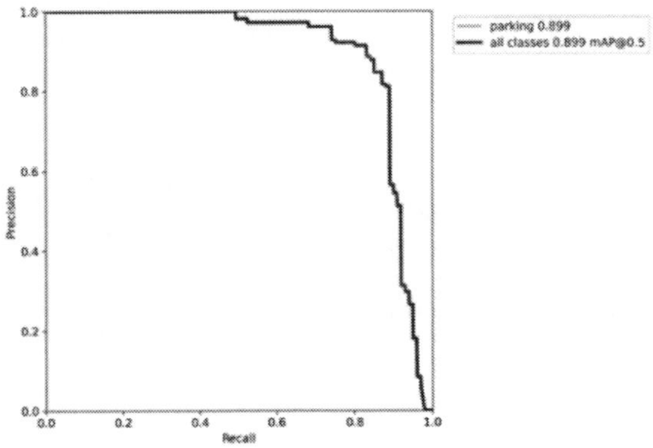

Figure 2. YOLOv7x Precision-Recall Curve.

(a)

(b)

Figure 3. a) YOLOv4 testing failed to identify some parking spots; b) YOLOv7x successfully detected parking spaces.

spaces, rather than the precision of the bounding box placement, we can infer that the model is highly effective at identifying available parking spaces in the given frame.

Fig. 4a shows an example where the YOLOv4 model detects a parking spot twice, unlike YOLOv7x (see Fig. 4b) where the model correctly detects the available parking spaces as one.

IV. CONCLUSION

In this paper, we proposed a new and innovative real-time street parking detection scheme that is based on the latest YOLO architecture, namely YOLOv7x. The network was trained on a new dataset mainly captured by our team and was designed to receive video input from a car mounted camera. We labeled parking spots that are within a distance of 5 meters from the car, and only focused on the right side of the street. This approach helps eliminate double counting parking spots and more accurately determining if the parking space is long enough to fit a car. Performance evaluations have shown that the YOLOv7x model outperforms the state-of-the-art YOLOv4 based approach in terms of both accuracy and detection. The performance of our model could be significantly improved by increasing the size and variety of our dataset. Future work will include new motion detection techniques that calculate how much a frame has shifted using global motion vectors to help analyze how many frames should be skipped after detecting a parking spot to find the next processing frame that has a parking spot. Additionally, we plan to add a separate network for detecting parking signs to provide a comprehensive solution that can be integrated into smart city infrastructures.

(a)

(b)

Figure 4. a) YOLOv4 detecting a parking spot more than once. b) YOLOv7x accurately detecting a single parking spot.

ACKNOWLEDGMENT

This work was supported in part by the Natural Sciences and Engineering Research Council of Canada (NSERC – PG 11R12450), and TELUS (PG 11R10321).

REFERENCES

[1] R. Arnott and E. Inci, "An integrated model of downtown parking and traffic congestion," Journal of Urban Economics, vol. 60, no. 3, pp. 418–442, 2006.

[2] J. Polak and P. Vythoulkas, "An assessment of the state-of-the-art in the modelling of parking behaviour," TSU REF, vol. 752, p. 31, 1993.

[3] MD. Meyer and M. McShane, "Parking policy and downtown economic development," J Urban Plan Dev ASCE, vol.109, no.1, pp. 27-43, 1983.

[4] V. Paidi, H. Fleyeh, J. H˚akansson, and R. G. Nyberg, "Smart parking sensors, technologies and applications for open parking lots: a review," IET Intelligent Transport Systems, vol. 12, no. 8, pp. 735–741, 2018.

[5] D. G. Chatmana and M. Manville, "Theory versus implementation in congestion-priced parking: An evaluation of SFpark, 2011–2012," Research in Transportation Economics, vol.44, pp. 52-60, 2014.

[6] Y. Shao, P. Chen, and T. Cao, "A Grid Projection Method Based on Ultrasonic Sensor for Parking Space Detection," IGARSS 2018 - 2018 IEEE International Geoscience and Remote Sensing Symposium, Valencia, Spain, pp. 3378-3381, 2018.

[7] E. Sifuentes, O. Casas, and R. Pallas-Areny, "Wireless Magnetic Sensor Node for Vehicle Detection with Optical Wake-Up," in IEEE Sensors Journal, vol. 11, no. 8, pp. 1669-1676, Aug. 2011.

[8] J. Chinrungrueng, U. Sunantachaikul, and S. Triamlumlerd, "Smart parking: An application of optical wireless sensor network," in Proc. Int. Symp. Appl. Internet Workshops, p. 66, 2007.

[9] X. Sevillano, E. Màrmol, and V. Fernandez-Arguedas, "Towards smart traffic management systems: Vacant on-street parking spot detection based on video analytics." in 17th International Conference on Information Fusion (FUSION), pp. 1-8, 2014.

[10] B. Sairam, A. Agrawal, G. Krishna, and S. P. Sahu, "Automated Vehicle Parking Slot Detection System Using Deep Learning," *2020 Fourth* International Conference on Computing Methodologies and Communication (ICCMC), Erode, India, pp. 750-755. 2020.

[11] T. Bazzaza *et al.*, "Automatic Street Parking Space Detection Using Visual Information and Convolutional Neural Networks," *2022 IEEE International Conference on Consumer Electronics (ICCE)*, Las Vegas, NV, USA, pp. 01-02, 2022.

[12] FFmpeg Developers. (2016). ffmpeg tool (Version be1d324) [Software]. Available from http://ffmpeg.org [retrieved February, 2023].

[13] OpenVINO Toolkit, 2020. GitHub repository. Available from https://github.com/openvinotoolkit/cvat. [retrieved February, 2023].

[14] AlexeyAB version of Darknet, 2020. GitHub repository. Available from https://github.com/AlexeyAB/darknet. [retrieved January, 2023].

[15] C.-Y. Wang and A. AB implementation of YOLOv7. GitHub repository. Available from https://github.com/WongKinYiu/yolov7 [retrieved February, 2023].

[16] G. Boesch, "YOLOv7: The Most Powerful Object Detection Algorithm (2023 Guide)," Available from https://viso.ai/deep-learning/yolov7-guide/. https://viso.ai/deep-learning/yolov7-guide/ [retrieved February, 2023].

[17] G. Jocher ultralytics version of YOLOv5. GitHub repository. Available from https://github.com/ultralytics/yolov5. [retrieved February, 2023]

[18] Compute Canada state-of-the-art advanced research computing network. Available from: https://www.computecanada.ca [retrieved February, 2023].

Approach to Modeling Energy Consumption of Auxiliary Consumers in Electric Vehicles

Tuyen Nguyen, Reiner Kriesten
Institute of Energy Efficient Mobility
University of Applied Sciences
Karlsruhe, Germany
e-mail: {Tuyen.Nguyen, Reiner.Kriesten}@h-ka.de

Hendrik Schulte
Mechanical Engineering and Mechatronics
University of Applied Sciences
Karlsruhe, Germany
e-mail: schultehendrik@aol.de

***Abstract*—To fully use the benefits of a hybrid energy storage system in an Electric Vehicle (EV), it requires an effective energy management system to control the energy flow. Knowledge of how energy consumption is generated in EVs is, therefore, important for the development of such a system. Since one of the biggest energy consumptions in modern EVs are generated by the auxiliary consumers, an auxiliary model needs to consider all relevant auxiliary consumers, including the heating/air-conditioning system and other switched, dynamic, comfort and continuous consumers. The vehicle interior is, therefore, modeled using a 1-zone air model and the auxiliary consumers (such as heating/air conditioning, power steering and others) are adjusted based on environmental conditions. The electric power steering's energy demand is calculated dynamically during the journey. The energy consumption of the heating system matches literature data with deviation less than 8%, while the consumption of other auxiliary consumers matches measured values with deviation of 2.8%. The focus is on the energy consumption of auxiliary consumers in EVs, which are a significant factor in the energy flows of the vehicle. The model accounts for factors, such as weather and driving behavior, that affect the use of auxiliary consumers and the resulting energy consumption.**

Keywords-Electric Vehicle (EV); Auxiliary Consumers; Heating, Ventilation, Air Conditioning (HVAC); Energy Consumption; Simulation Based; Weather.

I. INTRODUCTION

The utilization of Hybrid Energy Storage Systems (HESS) in electric vehicles has been the subject of growing interest in recent years, as it offers a unique combination of the benefits of both lithium-ion batteries and supercapacitors. .The integration of these two energy storage technologies results in a system that boasts both high energy density and high-power density. The effective management of energy flow between these two storage technologies requires the implementation of an intelligent Energy Management System (EMS) [12]. For the development and testing of such systems, it is necessary to model the energy flows in the vehicle as realistically as possible. Based on these models, the EMS can make a prediction of the expected range and suggest and/or take appropriate corrective steps if any action is required [12]. Alongside the powertrain, the auxiliary consumers are the second-largest energy consumers in an electric vehicle. Heating and air conditioning of the interior in particular consume a lot of energy [11]. Since electric vehicles have to operate with a limited energy capacity or comparatively long charging times, an accurate representation of the energy flows and the respective energy consumption is significant for model-based development approaches of management systems like the EMS presented in [12]. The use of auxiliary consumers in a vehicle depends on many different influencing factors. For example, the vehicle's lights depend on the weather, the time of day, the season, and the route. The use of the radio and the heating/air conditioning system is in return dependent on the driver. So, the presented model has to simulate the usage of the auxiliary consumers and the resulting energy consumption realistically, while at the same time taking external influences such as weather and route conditions into account.

Section 2 presents an investigation about the state of the art and gives an overview about modelling of auxiliary consumers in EVs. In Section 3, the individual models (vehicle cabin, Heating, Ventilation, Air Conditioning (HVAC) system, and other auxiliary consumers) are presented. Section 4 presents the individual validations of the used model in this project. Afterwards, Section 4 discusses and concludes the results of this work.

II. STATE OF THE ART

The paper of Basler [1], Baumgart [2], Konz et. al. [10], Kruppok [11] and Suchaneck [16] are studied to compare with the modeling approach of this investigation. The research focuses on modeling the HVAC system, as it is the main auxiliary consumer of an electric vehicle. Modelling the other auxiliary consumers via a constant power is sufficient since the transient behavior is not the focus, only their energy consumption. However, the additional consumers must be modeled dynamically based on ambient and route conditions for a realistic estimation of energy flows. Route-specific data is required and can be obtained from suitable sources or entered manually.

This work's focus is on energy flow calculations and does not aim to calculate additional range reductions of an electric vehicle or model comfort behavior. Instead, it aims to realistically calculate energy consumption through a realistic integration of route and environmental parameters into the auxiliary consumer model.

III. MODEL ENVIRONMENT

The model is generated in C-Code which is integrated into the simulation tool CarMaker. On one side through the interface, it is possible to set options and starting conditions for the model. On the other side, it is possible to return the calculated energy flows to the CarMaker environment for further processing [9]. While the transient behavior of the HVAC system determines the energy consumption significantly, the transient behavior of the other consumer (e.g., lightning, wipers) is not relevant for an accurate energy flow calculation. That's why the model is split into the blocks of HVAC and other consumers. The other consumers can be modelled as consumers with constant power demand when turned on.

A. *Heat Flow Balance of the vehicle cabin*

The vehicle's exterior surfaces are in heat exchange with the environment. Solar radiation passes directly through the glazed surfaces of the vehicle into the passenger compartment ($\dot{Q}_{trans}$) and the exterior surfaces of the body heat up. At low ambient temperatures, the heated body parts release their heat ($\dot{Q}_{body}$). A heat flow is supplied by the HVAC module ($\dot{Q}_{HV\,AC}$). Since the principle of mass conservation applies within the vehicle interior, the air mass flow supplied by the air-conditioning system must be able to escape again from the vehicle, as otherwise there would be an increase in pressure in the cabin, which is not the case (for real vehicles). The calculation of the dissipated heat flows $\dot{Q}_{loss}$ and the outgoing mass flow is shown in [7]. If the passenger compartment is considered as a holistic system, the following heat flow balance results:

$$\dot{Q}_{HV\,AC} + \dot{Q}_{loss} + \dot{Q}_{body} + \dot{Q}_{trans} = 0 \quad (1)$$

The heat flow from the HVAC system $\dot{Q}_{HV\,AC}$ can be calculated with the air mass flow $\dot{m}_{air}$, the specific heat capacity $c_{p,air}$ and the temperature difference ΔT.

$$\dot{Q}_{HV\,AC} = \dot{m}_{air} * c_{p,air} * \Delta T \quad (2)$$

The other heat fluxes, like convection, radiation and heat conduction, are calculated as described in [19].

B. *Influence of Solar Radiation*

The solar radiation entering the vehicle interior through the glazing heats up the interior of the vehicle, which additional heats up the air inside the vehicle cabin. As described in [14] it is possible to determine the sun's position through the day depending on the rotation around the north axis. The angle between the sun and the north axis is called azimuth (α_{sun}). To calculate which masses and which exterior surfaces of the vehicle cabin are heated up by the sun, it is necessary to calculate the relative angle between the sun and the vehicle. Figure 1 shows the relative rotation between the sun and the vehicle. The angle between the vehicle and the sun Δα is the difference between

$$\Delta\alpha = \alpha_{vehicle} - \alpha_{sun} \quad (3)$$

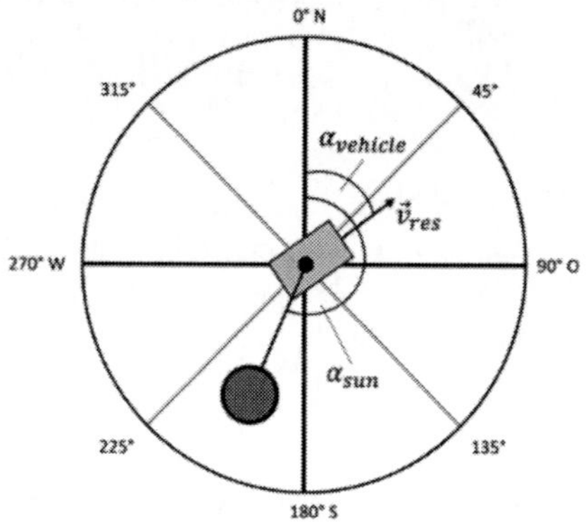

Figure 1. Relative rotation between a vehicle and the sun.

From Δα can be derived which surface and which window is heated up by direct solar radiation. If Δα is equal to zero or 180° the sun is directly in front or back of the vehicle and these are the only surfaces which are shined. If Δα is not equal to zero more than one reference surface is irradiated and depending on the value of Δα the shined surfaces are calculated in a look-up-table. The sun position is calculated as described in [14] and the solar intensity depending on date and time can be provided for different regions worldwide. For this paper, we will use Karlsruhe in Germany as an example, data provided by [4].

C. *Model of the vehicle cabin*

The vehicle interior is modelled as an air volume with a homogeneous air density and temperature distribution. The interior temperature depends on the incoming heat fluxes from the interior walls of the vehicle cabins and the incoming solar radiation. Furthermore, the interior temperature is increased, reduced or kept constant by the heating/air-conditioning system. The design of the vehicle interior model is based on the model by [11] but is adapted with regard to the model's own requirements.

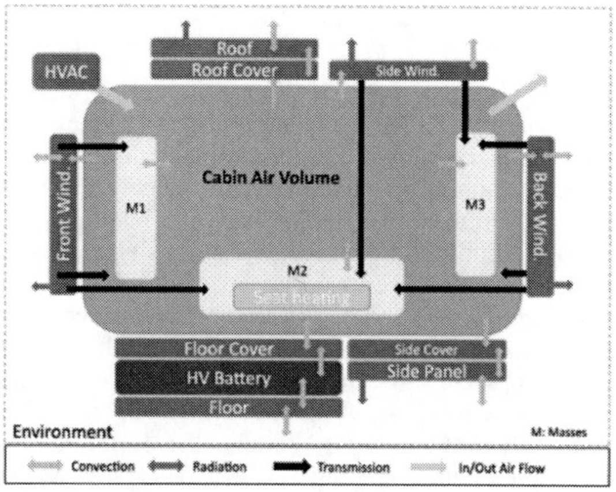

Figure 2. Design of the vehicle cabin model. Displayed are all modelled surfaces, masses and heat exchange mechanisms.

Figure 2 shows the design of the vehicle cabin model and all heat exchanging processes that appear in the vehicle cabin.

The external surfaces of the vehicle are in heat exchange via radiation (red arrows) and forced convection (yellow arrows). There are three glass surfaces in the vehicle cabin: Front window, side windows and back window. Through the glass surfaces, solar radiation is emitted into the vehicle cabin (black arrows). All incoming and outgoing heat flows are balanced via the cabin air volume. The temperature of the cabin air volume can be calculated with [1][11].

$$\Delta T = \int_0^t \frac{\dot{Q}_{air}}{m \cdot c_{p,air}} dt \tag{4}$$

Using (4), it is possible to calculate the surface temperatures of the vehicle cabin. For the calculation of the heat exchange coefficient, the vehicle cabin is simplified as a cylinder [2] and calculation of the coefficient is done with the methods from [19]. The solar radiation hitting the inclined surfaces $E_{dir,tor}$ can be calculated with the help of the direct horizontal solar radiation $E_{dir,hor}$ from the data from [4], the torsion angle of the surface θ and the solar elevation γ_{sun} [14]:

$$E_{dir,tor} = E_{dir,hor} * \frac{\theta}{\gamma_{sun}} \tag{5}$$

For a detailed calculation of θ see [14]. The diffuse solar radiation $E_{diff,tor}$ can be estimated with an isotropic approach using the horizontal diffuse radiation from the data $E_{diff,hor}$ and the rotation angle of the surface γ_{surf}.

$$E_{diff,tor} = E_{diff,hor} * 0.5 * (1 + \cos \gamma_{surf}) \tag{6}$$

The calculation of a projected surface, as in [7], is dispensed with in favor of the model complexity. The fraction of radiation which enters the vehicle cabin through the glazed surfaces is determined by the transmission coefficient τ. The radiation entering the cabin is calculated as a product of the sum of the diffuse and direct radiation times τ and the surface area A.

$$Q_{through} = (E_{diff,tor} + E_{dir,tor}) * \tau * A \tag{7}$$

It is assumed that the passengers are sitting and remain calm which results in heat flow from the passengers of $58 \frac{W}{m^2}$ [15]. Depending on the height and weight of the passenger the surface area of the passenger can be calculated with the Mosteller equation, which can be found in [18].

D. Controlling of the HVAC-System

The HVAC system can be operated in three different operational modes. In the preset mode, it is possible to set the desired cabin temperature and the inlet mass flow. The maximum mass flow of the HVAC system is in this case $9 \frac{kg}{min}$. The Min/Max - Mode aims to heat/cool down the vehicle cabin as fast as possible. If the temperature difference between the set temperature and the actual temperature is smaller than 3 K, the HVAC power is reduced. The automatic mode is modelled using the comfort perception of an average passenger. The comfort perception is described in [3] and can be modelled with the equations of [2]. The comfort perception is depending on the environment temperature, see Figure 3. Independent from the operation mode, the inlet air temperature is controlled by a PI - Controller. A Proportional–Integral–Derivative (PID)-controller has no significant advantage compared to a Proportional–Integral (PI)-controller because of the big thermal internal of the vehicle cabin [11]. A PI-controller is also used in the model of [13] to control the Air Conditioning (AC)-compressor. The power demand of the air conditioning is calculated using the Coefficient of Power (COP). The data comes from a comparable air conditioning system [10]. Using (8), the electric power demand P_{el} can be calculated.

$$COP = \frac{\dot{Q}_{HVAC}}{P_{el}} \tag{8}$$

The electric power demand when heating is calculated via a characteristic map using the output of the controller as input. The electric power demand is calculated as the output variable of the characteristic map. The characteristic map is constructed heuristically with data from [11] where the same HVAC system is modelled.

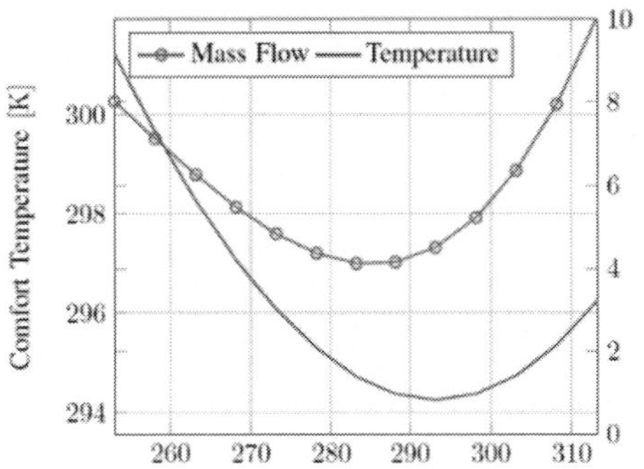

Figure 3. Comfort temperature and mass flow perceived as pleasant in the vehicle interior as a function of the ambient temperature. Calculated with the equatiosn from [2].

E. Modelling the other auxiliary consumers

Auxiliary consumers are divided into four categories: continuous (e.g., electronic control unit), comfort (e.g., radio and window heating), dynamic (electrical power steering), and switched consumers (can be switched by driver). Kruppok has published measured power values of the auxiliary consumers, which are used in this model, since the same electric vehicle is modelled [11].

The low beam in this model is switched on and off automatically. A distinction is made between summer and winter. In winter, the low beam is switched on automatically between 5 p.m. and 9 a.m. In summer mode, the period is shortened to the time between 8 p.m. and 6 a.m. From visibility of less than 150 m, the low beam is also switched on.

If the low beam is switched off, the daytime running lights are switched on as required by the Road Traffic Act [17]. The indicators are turned on when the internal CarMaker variable Vehicle on Junction is set and the steering angle is greater than 114.6°. The brake lights are turned on when the internal CarMaker Variable Brake is set. The wipers are used depending on the rain rate set before the simulation in CarMaker. The operation level is set automatically in the model. The Electric Power Steering (EPS) is modelled differently. Due to the highly dynamic power demand, the actual EPS power is calculated by a characteristic map.

IV. VALIDATION

The validation of the model includes, on the one hand, the validation of the heating/climate system and the associated thermal processes in the vehicle interior, and secondly the power requirements of the auxiliary consumers. Measurement data from the literature are used for validation. There are measurement studies of the temperature behavior of the vehicle cabin, which can be found in [6] and [11].

A. Validation of the Vehicle Cabin Model

The An essential component of the overall model with considerable influence on the temperature development in the interior is the vehicle cabin model. For the validation of the vehicle cabin model, the heating behavior of the interior is observed without air conditioning, without solar radiation and without wind. The measurement data used is taken from the work of Kruppok [11]. The test vehicle is pre-tempered to 275.5 K for a sufficiently long time, so that it can be assumed that each component has the same temperature. Subsequently, the vehicle is transferred to an environment with an ambient temperature of 291 K and the heating behavior of the interior is measured. The temperature is measured at different places inside the vehicle cabin and is averaged for validation. The vehicle is stationary and there is no irradiation from the sun. The same is also simulated in the model. To describe how well the model reproduces the experimental data, the coefficient of determination R2 is shown in (9). A detailed derivation of the coefficient of determination can be found in [5].

$$R^2 = 1 - \frac{\sum_{i=1}^{n}(y_i - \hat{y}_i)^2}{\sum_{i=1}^{n}(y_i - \bar{y}_i)^2} \qquad (9)$$

The coefficient of determination lies between $0 \leq R^2 \leq 1$. A value of zero indicates that there is no correlation between the measured values and the model, whereas a value of 1 can only be achieved if the measured values also describe the model at the same time. For the validation of the vehicle cabin model, the coefficient of determination between the heating curve of the model and the measured values is calculated. Optimal is a value close to one. By adjusting the heat transfer coefficients of the vehicle cabin model, a coefficient of determination of approx. 91.1% is achieved. Figure 4 shows that the heating of the interior is faster in the model than in the measurement, but the same final temperature is reached in both cases. For the determination of the energy demand of the air conditioner without taking comfort aspects, the interior model is considered to be sufficiently accurate.

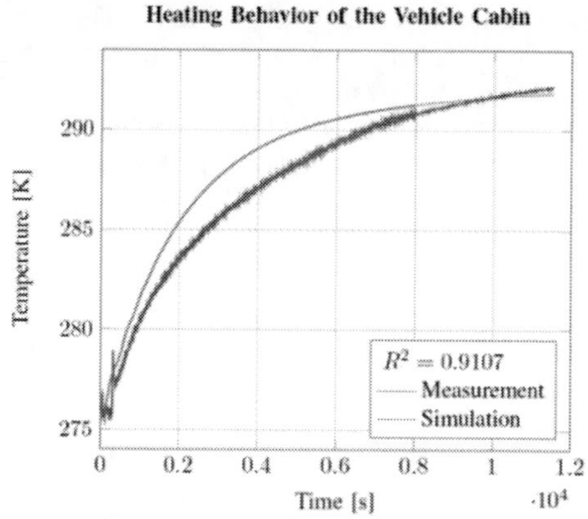

Figure 4. Heating of interior without air condit. with pre-heated interior.

B. Validation of the HVAC

Investigations by Kruppok have shown that the required inflow temperature of the air flowing into the vehicle, cabin is not achieved immediately after the heating/air-conditioning system is switched on [11]. Due to the thermal inertia of the system, it takes approx. 1000s (about 17min) until a stationary final temperature is reached. The behavior of the inflow temperature can be approximated by a transfer element with a second-order delay (PT2 element). The result is a well-fitting temperature curve in the model compared to the measurement data, see Figure 5.

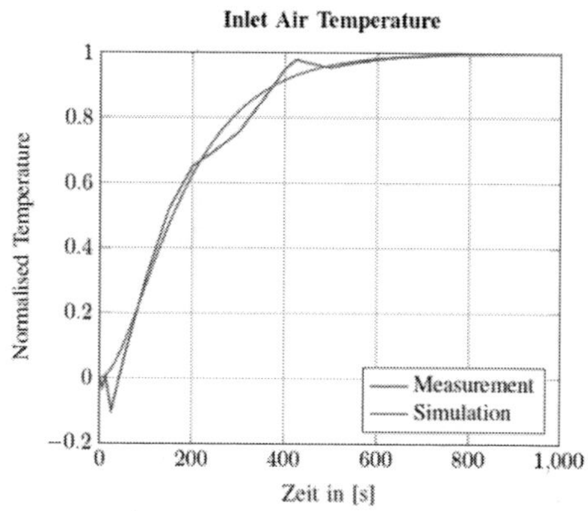

Figure 5. Heating of the interior without air conditioning with pre-heated interior.

It is assumed that the same behavior occurs in the heating mode because the HVAC is operated as a heat pump when in heating mode. So, the same components are in operation. By validating the heating behavior of the vehicle cabin model, the entire HVAC system, but especially the controller setting can be validated. As a reference, a measurement in which the interior temperature is recorded with active heating. This is the same as the average temperature, determined in the model. The test vehicle is pre-conditioned to 266.2 K and the target temperature of the interior is 297.2 K. The measurement data is published in [6]. In the preset mode, the temperature behaves like the real-world measurement. The heating system is operated with the maximum mass flow of $9\,\frac{kg}{min}$.

A similar mass flow is also used for the measurement, due to the low ambient temperature and the preconditioning of the vehicle. In the real vehicle, there is a clear overshoot of the interior temperature. This is avoided in the model by controller setting. The target temperature is reached at the same time after approx. 800s. The steady-state final temperature of the simulation is a few tenths of a Kelvin higher than the setpoint temperature, but this deviation reduces in the further course of the simulation. In min/max mode, the setpoint temperature is reached much earlier, as desired. A stationary final state is already reached after approx. 400s. This is achieved by a maximum temperature of the incoming air and a maximum air mass flow. The validation of the heating-up behavior shows that the model can be used to simulate the real vehicle interior very well. Through the different operating modes, it is possible to simulate different scenarios.

C. *Validation of The Energy Consumption*

In his work, Gutenkunst carried out five test drives of Bruchsal - Karlsruhe - Bruchsal and measured the distance travelled and the energy consumption [8]. Those measured data is used as a reference for the results of the simulation. In the simulation environment, the model has calculated with the air conditioning switched off, in good visibility, without rain and during the day. During the simulation, the local speed limits along the route are adhered to exactly. Table I shows the results of the measurement runs and the simulation.

TABLE I. COMPARISON ENERGY CONSUMPTION. MEASUREMENT DATA ARE TAKEN FROM [8].

Number	Distance [km]	Measurment Energy [kWh]	Measurment Consumption [kWh/100km]	Simulation Energy [kWh]	Simulation Consumption [kWh/100km]
1	56.92	7.77	13.65		
2	56.87	9.72	17.08		
3	56.29	9.95	17.62		
4	57.24	9.44	16.50		
5	57.62	10.17	17.65		
⌀	56.99	9.41	16.50	9.67	17.31
Deviation of ⌀				**2.78 %**	**4.89 %**

The normalized energy consumption in the simulation is approx. 5% higher than the average consumption of the measurement runs. This is due to the following reasons: The route length in the simulation already deviates 1.94% from the average real route length. The reason for this is the inaccuracy of GPS data, with which the route was created in CarMaker. Since the normalized consumption results from the total energy demand and the route, there are deviations. The pure energy demand is also higher in the simulation despite the shorter route. The reason for this is the unknown use of the auxiliary consumers and a possible velocity deviation during the measurement. Since the deviation in the simulation to the measurement is less than 3%, it can be assumed that the general calculation of the energy consumption is sufficiently accurate. The data from measurements two to four are done with normal traffic. Since the traffic is not modelled in the simulation the comparison between measurements two to four and the simulation are expected to be better. The comparison shows a deviation of only 0.33% for the comparison of total energy consumption. The normalized consumption is now only 1.4% higher than in the measurement.

D. *Validation*

The energy consumption of the HVAC system in the heating case is validated with the data from [11]. Since a similar HVAC model for the same reference, the vehicle is described in [11], the comparison is legitimate. The power demand in the heating case in the model of the present work is determined via a characteristic map. The map is developed heuristically. The input variable of the characteristic map is the output of the controller, which describes the extent to which the temperature of the incoming air must be increased by the heating system. The output variable is the electrical power in Watts. The energy consumption calculation is carried out in the time frame of the WLTP cycle. Figure 6 shows the result of the validation with different ambient and target temperatures in the vehicle cabin.

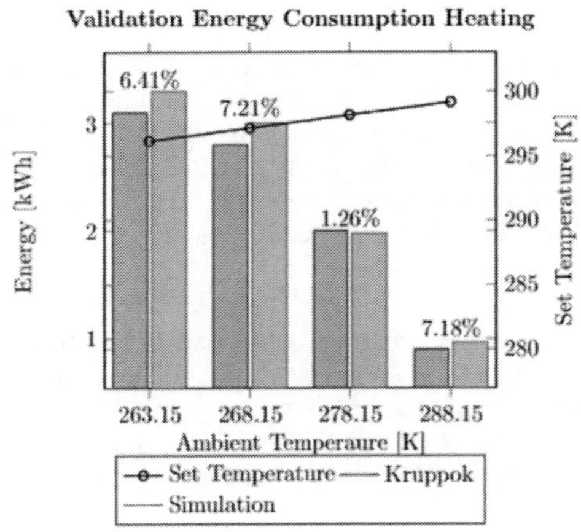

Figure 6. Validation of the energy consumption in the heating case at different ambient temperatures in the WLTP cycle. Comparison data are taken from [11].

The percentage values above the bars describe the relative deviation between the simulation and the data from [11]. The approximation of the power demand of the heating system via a map provides sufficiently accurate matches to the data from the literature. The relative deviation is in all cases smaller than 7.25%. At low ambient temperatures, however, the energy consumption in the simulation is higher than in [11]. One possible reason for this is deviating heat transfer coefficients, which lead to increased heat loss from the vehicle cabin at low ambient temperatures. The heat transfer coefficients in [11] are not fully known.

V. DISCUSSION AND CONCLUSION

The validation shows a good result if the model is compared with data from the literature. The deviation is relatively small and is, therefore, acceptable for the objective of this work. The model construction of the HVAC system is based on similar models, which can be found in literature, e.g., [1],[11] or [16]. The aim was to build a simpler model with at least nearly good simulation results as other, more complex models. This is achieved but only validated for the heating case. In addition, the calculation of the power demand of the heating system with a characteristic map is not very accurate although it works well in this case. A detailed measurement of real currents would bring more accurate information into the model. But overall, the investigation has shown, that the model results are quite accurate for the cases investigated. With more detailed data from the reference vehicle, especially for cooling cases and the explicit use of different auxiliary consumers the model can be updated to a more detailed model with a highly accurate energy consumption calculation.

The auxiliary consumer model makes it possible to determine the energy consumption of all the auxiliary relevant in an electric vehicle. The model shows good agreement with measured data which can be found in the literature, despite its simplifications and assumptions and with considerably less modelling effort. The deviation of the energy consumption of the heating system to values from the literature is less than 8% and the consumption of the other auxiliary consumers deviates only 0.33% from the measured values. At the same time, the structure of the model makes it possible to link the model to the simulation tool CarMaker. Via the interface between the model and CarMaker, it is possible to connect other models, such as an EMS, to the auxiliary consumer model. This work has shown that even a less detailed model can produce similarly accurate results about the energy consumption of the auxiliary consumers than detailed models.

ACKNOWLEDGMENT

This work was carried out as part of VEHICLE project, sponsored by INTERREG V A Upper Rhine Programme - Der Oberrhein wächst zusammen: mit jedem Projekt, European Regional Development Fund (ERDF) and Franco-German regional funds (Baden-Württemberg, Rhineland-Palatinate and Grand Est).

REFERENCES

[1] A. Basler, "Eine modulare Funktionsarchitektur zur Umsetzung einer gesamtheitlichen Betriebsstrategie für Elektrofahrzeuge, [in English: A Modular Functional Architecture for the Implementation of a Holistic Operating Strategy for Electric Vehicles]", KIT Scientific Publishing, Karlsruhe, 2015.

[2] R. Baumgart, "Reduzierung des Kraftstoffverbrauches durch Optimierung von Pkw-Klimaanlagen, [in English: Reduction of fuel consumption through optimization of vehicle air conditioning systems]", Chemnitz, Techn. Univ., Diss., 2010. Publ. sci. scripts, Auerbach, 2010, ISBN:9783942267014.

[3] U. Deh, "Kfz-Klimaanlagen, [in English: Automotive air conditioning]", Service Fibel. Vogel Business Media, Würzburg, 3rd edition, 2011, ISBN:978-3- 8343-3212-7.

[4] Deutscher Wetterdienst - Ortsgenaue Testreferenzjahre von Ortsgenaue Testreferenzjahre von Deutschland für mittlere, extreme und zukünftige Witterungsverhältnisse: Handbuch, [in English: German Weather Service - Locally accurate test reference years of Locally accurate test reference years of Germany for average, extreme, and future weather conditions: Handbook], Offenbach, 2017.

[5] L. Fahrmeir, C. Heumann, R. Künstler, I. Pigeot, and G. Tutz, "Statistik: Der Weg zur Datenanalyse, [in English: Statistics: The way to data analysis]", Springer-Textbook, Springer Berlin Heidelberg, Berlin, Heidelberg, 2016, ISBN:978- 3-662-50371-3.

[6] T. Fleck, "Wer heizt am besten?, [in English: Who heats the best?]", Motor Press Stuttgart GmbH & Co. KG, Stuttgart, 2015.

[7] H. Großmann, "Pkw-Klimatisierung: Physikalische Grundlagen und Technische Umsetzung, [in English: Car air conditioning: physical principles and technical implementation]", VDI-Book Ser. Springer Berlin / Heidelberg, Berlin, Heidelberg, 2nd edition, 2013, ISBN:978-3-642-39841-4.

[8] C. Gutenkunst, "Prädiktive Routenenergieberechnung eines Elektrofahrzeugs, [in English: Predictive route energy calculation of an electric vehicle]", PhD thesis, Karlsruher Institute for Technology, 2020.

[9] IPG Automotive Group, "Reference Manual," Version 9.1.1, CarMaker, Karlsruhe, 2020.

[10] M. Konz, N. Lemke, S. Försterling and M. Eghtessad, "Spezifische Anforderungen an das Heiz-Klimasystem elektromotorisch angetriebener Fahrzeuge, [in English: Specific requirements for the heating/air-conditioning system of electric motor-driven vehicles]", volume 233, FAT-publication series, 2011, ISSN:2192-7863

[11] K. Kruppok, "Analyse der Energieeinsparpotenziale zur bedarfsgerechten Reichweitenerhöhung von Elektrofahrzeugen, [in English: Analysis of energy saving potentials for increasing the range of electric vehicles in a demand-oriented approach]", expert publisher GmbH, Tübingen, 1st edition, 2020, ISBN:978-3-8169-3516-2.

[12] T. Nguyen, R. Kriesten, and D. Chrenko, "Concept for generating energy demand in electric vehicles with a model based approach," MDPI - Applied Sciences, 2022, doi:10.3390/app12083968.

[13] F. Nielsen, S. Gullman, F. Wallin, A. Uddheim, and J.-O. Dalenbäck, "Simulation of energy used for vehicle interior climate," SAE International Journal of Passenger Cars - Mechanical Systems, 2015, doi:10.4271/2015-01-9116.

[14] V. Quaschning, "Regenerative Energiesysteme: Technologie, Berechnung, Simulation; mit 113 Tabellen, [in English: Renewable energy systems: technology, calculation, simulation; with 113 tables]", Hanser, Munich, 7th edition, 2011, ISBN:978-3-446-42944-4.

[15] C. Schmid, T. Baumgartner, C. Bucher, J. Nipkow, and C. Vogt, "Heizung/L¨uftung/Elektrizität: Energietechnik im Gebäude, [in English: Heating/ventilation/electricity: energy technology in buildings]", Construction & Energy, vdf University Press AG at the ETH Zurich, Zurich, 6th edition, 2020, ISBN:978- 3-7281-4020-3.

[16] A. Suchaneck, "Energiemanagement-Strategien für batterieelektrische Fahrzeuge, [in English: Energy management strategies for battery electric vehicles]", Karlsruher Institute for Technology, Karlsruhe, 2018.

[17] R. Süßbier, "Hinweise zum korrekten Umbau von Beleuchtungseinrichtungen, [in English: Information on the proper conversion of lighting equipment]", GTÜ, Stuttgart, 2014.

[18] T. Vu, "Standardization of body surface area calculations," Journal of Oncology Pharmacy Practice, 8(2-3) pp. 49–54, 2002.

[19] VDI e.V., "VDI-Wärmeatlas, [in English: VDI Thermal Atlas]", Springer Vieweg, Berlin, Heidelberg, 2013, ISBN:978-3-642-19980-6.

A Holistic Approach on Automotive Cybersecurity for Suppliers

Jose Ángel Gumiel Quintana
Electronics and Communications Unit
Fundación Tekniker
Eibar, Spain
e-mail: jagumiel@tekniker.es

Jon Mabe Álvarez
Electronics and Communications Unit
Fundación Tekniker
Eibar, Spain
e-mail: jmabe@tekniker.es

Jaime Jiménez Verde
Electronic Technology Department
University of the Basque Country (UPV/EHU)
Bilbao, Spain
e-mail: jaime.jimenez@ehu.eus

Jon Barruetabeña Pujana
Mechatronics Area
BATZ Group
Igorre, Spain
e-mail: jbarruetabena@batz.es

Abstract— **Electronics are increasingly present in automobiles. This has led to a change and automotive parts suppliers are forced to integrate electronic components into their mechanical parts. As if this were not enough of a change, the transition to electronics brings with it other issues, such as cybersecurity. Vehicles are becoming increasingly technological and connected, and car manufacturers are already asking suppliers to ensure cybersecurity. For a traditional supplier, this is something new, and what they often ignore is that it affects not just the product, but the whole organization. Therefore, the purpose of this article is to shine a spotlight on cybersecurity and explain simply, but honestly, the new challenges they face. The automotive industry is a sector that has been able to reinvent itself throughout its history, and in these times of abrupt and accelerated change, it must do so again.**

Keywords - Automotive Engineering; Cybersecurity; Information Technologies; Connected Vehicle.

I. Introduction

Automobiles have already become computers on wheels. 90% of the technological innovation in vehicles is electronic [1], and the number of electronic elements in cars is predicted to continue to grow [2]. Currently, electronics account for 30% of the total cost of a vehicle, and are expected to account for 50% by 2030 [3].

This is a major disruption. The trend towards the introduction of electronics, software and smart elements is a great opportunity for technology companies, from giants to start-ups, as it allows them to enter a new market [4][5]. On the other hand, it is a threat to traditional suppliers, so-called TIER 1 suppliers, who are forced to renew themselves [4] or be relegated to commodity manufacturers, products with very low cost and little added value.

Many of these well-established suppliers are already aware of this situation and have begun to take steps to integrate electronics into their products [6]. They have some advantages over their new competitors in that they are experts in the design and manufacture of mechanical parts in a variety of materials, they have a good understanding of how the industry works, the timelines and processes for mass-producing parts and delivering them on time, and what is more, they have the prior confidence of the automobile manufacturers, also known as Original Equipment Manufacturers (OEMs).

The integration of hardware (HW) and software (SW) in mechatronic components is not a trivial matter, as it requires knowledge of electronic design and SW development, which is precisely what these companies often lack [7]. In addition, there are a number of well-established regulations that must be complied with [8], as well as new ones that are emerging due to the prevailing needs, such as cybersecurity [9].

Automotive electronic systems are connected to different communication buses and share data with each other, they are not isolated [10][11]. This brings new challenges, as secure communication must be ensured and data must not be compromised [12]. But security goes much further; it starts in the corporation, goes through the design and development of the system, passes through manufacturing, and reaches the vehicle, where it must be robust so as not to endanger other elements, and invulnerable to the passage of time and new technologies.

The purpose of this article is to show that security is a very broad issue, encompassing the whole organization and not just the system developers, and must be ensured throughout the vehicle´s lifetime. Section II explores the differences between safety and security, with particular emphasis on the role of ISO standards in improving both aspects of automotive systems. Section III delves deeper into the topic of cybersecurity, examining its impact on every stage of the automotive product lifecycle, from the office to the end of the vehicle's life. This section is divided into six subsections that examine specific areas of the automotive ecosystem, identifying vulnerabilities and suggesting protective measures. In Section IV, the consequences of potential cybersecurity threats involving outdated vehicles is discussed, examining some vulnerabilities and highlighting the importance of collaboration between OEMs and suppliers. Finally, Section V presents the conclusions of the research and outlines potential areas for future work on the topic.

II. SAFETY & SECURITY

Safety and security are two distinct, but closely related concepts [13]. While safety focuses on preventing accidents, injuries or fatalities through the design and operation of the vehicle, security aims to protect the vehicle and its occupants from unauthorized access, theft, or malicious attack. It cannot be ignored that a security breach could also have an impact on the functional safety of the vehicle, putting occupants and other road users at risk. Several standards have emerged to address the growing concern for vehicle safety and security. One is ISO 26262, which came out in 2011. This is about functional safety. In a nutshell, the standard helps to classify the electronic system according to the severity, exposure, and repeatability of its risks, and urges to take different measures, depending on the category of the system, to make it safe. The standard applies to the entire project lifecycle and includes activities and deliverables for documentation and traceability. In 2018 the second version of this standard was released, and already anticipated the security issue, including common methods for functional safety and cybersecurity [14] to ensure the protection of vehicles from malicious attackers. This already hinted at the concern for cybersecurity, and the close relationship it has with functional safety despite being different areas. Both are collaborative elements that require comprehensive system engineering [11].

On the other hand, and more recently, ISO/SAE 21434, which deals with cybersecurity, appeared in 2021. Analyzing the standard, it resembles a compendium of best practices, because cybersecurity is not something generic; each protocol, each electronic system and each SW module may require a different approach.

Unlike other standards, such as ISO 16750, where the tests to be performed and the pass ranges are detailed, these are not defined. A study must be conducted to analyze the system, and tests to validate and verify the operation of the system must be previously defined, both for functional safety and cybersecurity.

III. FOCUS ON CYBERSECURITY

Security refers to protecting critical system assets from threats and mitigating their impact on the system [11]. These assets can be anything of value, either to the company or to the final product. If vulnerabilities exist, they can be exploited by malicious users or attackers.

There has been a gradual process of digitization in companies [15]. For example, customer communications are often on-line, information generated is stored digitally in repositories and databases, and production control can be done remotely. At the same time, in-vehicle communications have evolved [16], users demand wireless connectivity for their devices and OEMs can remotely update their vehicles' SW. All this makes it clear that cybersecurity must cover the entire lifecycle of an automotive component, from the concept phase to the end of the vehicles' life.

Although it may appear to be a topic that has already been addressed at the academic level, it has been observed that TIER 1s are unaware of the implications and costs to their business of implementing the concept of cybersecurity [17]. In some cases, lack of knowledge and prejudice lead them to believe that it is something trivial and easy to implement. In addition, it has been observed that some OEMs have the same lack of knowledge and ask for fuzzy requirements in their Request for Quotation (RFQ), which can be very vague cybersecurity specifications [18] or exaggeratedly high for the type of product. To illustrate how cybersecurity encompasses not only the product, but the entire organization, development processes, and the entire product lifecycle, the following is a breakdown of the areas in which security must be ensured in an automotive company.

A. *Cyb-Sec at the Office*

Cybersecurity is a culture. Employees must be educated on the subject and embrace the fact that security starts with them. In the automotive industry, sensitive customer data is at stake [19]. Competition is fierce, and OEMs try to differentiate themselves from each other in terms of design, quality, and innovation. Therefore, confidentiality must be ensured, and a secure working environment must be in place.

It is not uncommon for companies in this sector to have access control at the entrance to their facilities. These systems allow access only to authorized personnel and keep a record of entry and exit times.

In the office, the Information and Communication Technology (ICT) department must ensure the security of the network infrastructure. Part of their job is to implement preventive measures to avoid unauthorized access, modification, deletion and theft of resources and data, including industrial espionage [20]. These security measures may include authentication, access control, application security, firewalls, Virtual Private Networks (VPNs), behavioral analysis, Intrusion Detection and Prevention Systems (IDPS) and wireless security.

The international standard ISO 27001 addresses this issue. It ensures the confidentiality and integrity of data and information, as well as of the systems that process them, allowing the organization to assess the risks and implement the necessary controls to mitigate or eliminate them. This may be combined with periodic internal audits.

There are some best practices that can help secure communications.

- **Network segmentation:** This is an effective way to prevent potential intruder exploits from spreading to other parts of the internal network. It is possible to create different subnetworks; some typical examples are a subnetwork for employees and another one for visitors and external devices (such as personal laptops or smartphones) or subnetworks for different workgroups in the organization [21].
- **Firewall:** It is a network security system that monitors and controls incoming and outgoing network traffic. It allows to establish a barrier between a trusted network and an untrusted network. By applying rules, a firewall can allow or deny incoming and outgoing traffic from different IP addresses, protocols and ports [21].
- **Demilitarized Zone:** Creating a demilitarized zone (DMZ) can also be a good idea. This is a subnetwork that lies between the public Internet and private networks. It allows the enterprise to access untrusted networks, while ensuring the security of its private Local Area Network (LAN). Services are exposed, but the middle layer protects sensitive data on the intranet with a firewall that filters traffic [22].
- **Network devices protection:** A primary way to improve network infrastructure security is to harden devices such as routers, access points, servers, etc. Measures can include restricting physical access and, in some cases, protecting them from threats such as fire or water. In the digital realm, secure user access should be ensured, strong administration passwords should be used, device configurations should be backed up and devices should be tested regularly.
- **Access to information:** The organization must provide a secure way to store and access information. There are confidential projects that contain data that should not be accessible to all employees [20]. An information system based on user permissions could prevent read and write access to sensitive documentation. Similarly, the use of version control systems, such as Git or SVN, can increase productivity while maintaining confidentiality.
- **VPN:** This is often the preferred solution to allow remote workers to establish a secure connection to the corporate network. It creates a secure tunnel between the remote worker's computer and the corporate network [23]. It can be used both for accessing company resources remotely as well as for protection when using public connections, as the traffic is encrypted [24]. The main security problem is if an intruder gains access to the virtual network. The knowledge of authentication protocols helps to solve this problem. Today, it is even possible to introduce a double authentication factor by receiving a unique and temporary key on the mobile phone to verify the authenticity of the person who wants to connect, thus rejecting imposters.
- **Updated software:** Keeping SW up to date is important. Developers work hard to maintain compatibility and fix bugs and security vulnerabilities.

Despite the efforts of the ICT department to secure the work environment, this is pointless if employees are not aware of the importance of cybersecurity. There are actions that are solely up to them, such as using strong passwords and renewing them regularly, being suspicious of emails containing hyperlinks or suspicious files or managing documents with appropriate backups, among others.

B. Cyb-Sec in the Development Phase

Automakers are already starting to hold TIER 1 suppliers accountable for cybersecurity. Although this is new, it will soon become a common requirement due to the integration of electronic systems in the car and connectivity. Therefore, HW, and SW engineers will have to develop the project with the concept of cybersecurity in mind.

The way of ISO/SAE 21434's breaks down development is similar to the functional safety standard in that there is still a concept phase and a product development phase. It also adds a section on operations and maintenance during the post-development phase, indicating that there may be incidents, corrections, and updates. Unlike ISO 26262, the cybersecurity standard does not distinguish between HW and SW but is understood as a system. The following are some of the tasks that must be performed when developing a new product.

1) Concept definition

This part details the requirements for the concept phase. Its main objectives are:

1. **Define the item, the operational environment, and its interaction with other items:**
   This is an initial activity in which a preliminary study and design of the architecture is carried out according to the description of the item, an analysis of the known interactions with other components and some assumptions about its operating environment.
2. **Specify cybersecurity goals and cybersecurity claims:**
   Next, an analysis of the item is conducted. Cybersecurity engineers must perform a Threat Analysis and Risk Assessment (TARA). This consists of a matrix to identify potential threats and their likely attack vector, classify them according to their characteristics and their impact on security, operational, privacy and financial issues, and obtain an impact level. It also evaluates the knowledge required by the attacker to execute the identified attack. Ultimately, the aim of this document is to define cybersecurity goals to make a system robust and reliable against hackers and intruders. For those familiar with ISO 26262, this will remind them of the

Hazard Analysis and Risk Assessment (HARA) document.

While a threat analysis focuses on how an attacker might exploit vulnerabilities to gain access to resources or sensitive data, threat modelling tries to identify potential threats to the item's ecosystem and its periphery, as well as any vulnerabilities that could be exploited by those threats. Performing this work is also encouraged.

3. **Specify cybersecurity requirements and allocate them to the item or to the operational environment:**

   Based on the above activities and once the threats and cybersecurity goals have been identified, the next step is to establish requirements to meet those targets. That is, what measures will be taken to mitigate the risks and protect the asset against potential attacks.

2) Product Development

During this phase, the cybersecurity specifications must be defined. For this purpose, the HW and SW architectures of the system must also be described, although these may be modified to meet the security goals.

At the HW level, details are needed such as the communication protocol that the component will use to connect to the vehicle or the role of the item. This is important because a point-to-point communication protocol does not have the same security measures as a multiplexed or wireless protocol. The same is true for the role; the measures change if the item is a master or a slave, as well as if it is only responding to requests or sending commands to other systems.

There are also HW components, such as the Trusted Platform Module (TPM) that ensure the authenticity of the device or that the component's firmware has not been tampered with by a third party. This enables SW integrity reporting and cryptographic key creation and management. The applications are varied, but it is used to have a secure identification between the ECU and the component, as well as to identify against the deployment of updates to the system. For example, if an attacker modifies the firmware, the key will change, and the component will no longer be recognized by the vehicle as a trusted system.

Implementing cybersecurity in SW can be done in several ways. e.g., SW design patterns such as modularity, abstraction or layering contribute to system robustness. Similarly, domain separation and process isolation are two measures that limit privilege escalation and access to certain data and resources, making it more difficult for an attacker to gain control of the system. Simplicity and minimization also play a critical role; the easier it is to use and communicate, the easier it is to detect vulnerabilities, and if the system has only the necessary features, without extras, it will be less vulnerable because fewer violations will go undetected.

When the electronic design is mainly based on sensor integration, it is also necessary to ensure that the sensors can be calibrated and that the EEPROM can be locked so that it cannot be tampered with. The same applies to other procurement elements.

3) Cybersecurity Validation

This is the process that validates through testing the assumptions that were made in the previous stages [11]. These tests consist of emulating or simulating cyberattacks and testing the effectiveness of the security measures implemented, thus validating their functioning.

To validate the design there are several types of tests. Some of them will be mentioned and briefly explained below.

- **Penetration test:** Evaluates an application's attack surface for potential SW weaknesses that, if left unaddressed, could lead to exploitable vulnerabilities. This could result in remote code exploitation or sensitive information exposure. The way is to take over the device or application. It is usually combined with a vulnerability scan of the perimeter, where the system is located.
- **Vulnerability scan:** It is a SW that detects certain vulnerabilities in a device, such as well-known public vulnerabilities and configuration errors that pose a high risk of compromise within a network or system. For example: Remote Code Execution (RCE), Data Exposure, Denial of Service (DoS) vulnerabilities, etc.
- **Security scan:** Checks for misconfiguration, such as unencrypted files, unpatched systems, inadequate firewall or use of weak cryptographic methods or suites.

There is another type of analysis that requires knowledge of the communication protocol used by the device. In this way, it is necessary to check that the communication cannot be interrupted, that the identity of the device or sensor cannot be impersonated and that there is no repudiation by other systems. For certain applications, it must also be ensured that the data is protected against attacks (such as man-in-the-middle, eavesdropping or spoofing) or that data cannot be manipulated [11].

4) Product Maintenance

Over time, new technologies and tools emerge, hackers acquire new knowledge and discover new vulnerabilities, and systems that were once secure can become susceptible to attacks.

Connectivity has enabled cars with Over-The-Air (OTA) updates, which not only improve the performance and functionality of a vehicle already on the road, but also make it possible to fix bugs and security breaches on the fly.

When a security vulnerability or bug is discovered in the SW, it should be fixed and securely, quickly, and seamlessly updated without the need to visit the vehicle maintenance shop or garage [11]. The supplier must now ensure that its part is secure and respond to any incidents. The upside is that, although they have an additional role in maintaining safety and security, in some cases they may be able to avoid a recall, because there are issues that could be resolved remotely with a SW update to the component.

Therefore, a secure online SW update is required for every ECU in the autonomous and connected vehicles [11].

C. Cyb-Sec at the Testing Department

Automotive suppliers' typically have departments or laboratories dedicated to testing and measuring parts, whether they are prototypes, pre-series, or final products. Physical security measures are in place to protect know-how and intellectual property. In most cases, the site is protected by walls that prevent visibility from the outside, as well as having access control so that only certain people within the organization are allowed to enter.

However, not all security is physical, but the know-how of the ICT department is also relevant here. Because of the information handled in these facilities, some security measures are taken into consideration. For example, the organization may provide operators with digital cameras, so that they do not use their connected smartphones to take pictures nor videos of the products or tests. This department may be under a different subnetwork to further control communications with the outside world [21]. When storing sensitive test and prototype information, it is essential to have a system for managing backups, which can be stored and encrypted.

Protecting this information is important. An attacker could use it for a variety of purposes, such as industrial espionage, plagiarism, dissemination of results or defamation. On the other hand, backups are also important because in some cases there is information that may be requested by the OEM to solve problems or improve parts.

D. Cyb-Sec at the Production Line

The production line is the place where parts are manufactured and assembled in series to be shipped to the customer. Nowadays the lines are highly automated and quality control is performed on every part that is produced.

In this case, cybersecurity must also be considered. In many projects, the electronics are engraved or finished on the line. The SW is loaded, the sensors are calibrated, the device is configured with some data and finally, the EEPROM is locked so that the electronics cannot be manipulated again by third parties.

A security breach on the production line could affect manufacturing in several ways. For example, an attacker could remotely access the operation of the machines, obtaining data and parameters, changing settings, and even stopping manufacturing. It could cause devices to be programmed incorrectly, or to write off parts that should be rejects.

This could impact business and customer relationships. For these reasons, the ICT department will need to control communication and access as it does in other areas of the organization. When data is recorded on devices, a subsequent check should be made to ensure that the recording is correct and, as a last step, the device should be locked so that it cannot be recorded again.

Usually, a record is kept of the parts that come out and their characteristics. It is in the interest to protect this data and store it properly, with the desired encryption measures.

E. Cyb-Sec in the Vehicle

After the entire design and manufacturing process, the final product reaches the OEM, who installs it in the vehicle and sells it to the public. If the analysis has been carried out well, the system should be secure and not pose a risk to other elements of the vehicle.

On the other hand, the OEM also has certain cybersecurity responsibilities, as it sends commands to different systems, has access ports to ECUs, such as OBD-II, and in some cases can collect information about the performance and operation of its vehicles [11]. Some measures can be applied by the carmaker, such as secure boot process, IDPS or communication encryption.

If vehicle usage data is collected, it must be anonymized, without driver information, and the transmission paths must be secure. Alfa Romeo is a pioneer in providing a complete history of car telemetry data, guaranteeing its authenticity thanks to Non-Fungible Token (NFT) technology and digital certificates. The first vehicle in using NFT and blockchain technology will reach the market in 2023. It will be able to record data on vehicle's manufacture, mileage, electric battery cycles, overhauls, part changes, etc. This will provide complete and tamper-proof traceability over the vehicle's lifetime [25].

F. Beyond the Vehicle

This subsection is intended to make the reader aware that safety goes beyond the digital and that poor design can compromise the integrity of the driver.

Today's cars are equipped with Advanced Driver Assistance Systems (ADAS). Some of them are based on computer vision and include in-vehicle video cameras, proximity sensors, RADAR, and LIDAR technologies.

These technologies are sometimes unobtrusive, but they are already being used in some vehicles to achieve a certain degree of semi-autonomous driving. It should be kept in mind that the environment in which the car moves, the real world, can also be hacked.

There are methods an attacker can use to provide wrong information to the vehicle. McAfee Advanced Threat Research conducted research that revealed these risks. Manipulating road signs with small stickers caused the car to misinterpret them. Similarly, using a magnet to place numbers on speed signs, it was possible to mislead the vehicle about the maximum speed of the road [26], posing a serious threat to occupants and road users. Another hack is the phantom attack [27], which consists of projecting images onto the road surface. Cars could be tricked into recognizing fake pedestrians or signals, and even non-existent lanes.

IV. The Aging of the Connected Vehicle: Cybersecurity Concerns

In 2022, the average age of the European car fleet was around 12 years [28]. Looking back a decade ago, cars have changed significantly, especially in terms of connectivity. Although vehicles then already had a lot of electronics, it was nothing compared to all the systems that are included today, and they lacked wireless connectivity to the outside world. As a result, there were fewer potential entry points for cyber threats, which made them inherently more secure. If the mechanics of the purchased vehicle were good and maintenance was adequate, an old vehicle could be in service for many years, well beyond its average useful life. In such a case, the driver would have to forgo the new safety features that a new car could provide.

More than 400 million connected vehicles are expected to be in use by 2025 [29], and each vehicle will produce 25GB of data per hour [30]. Several factors, including the widespread adoption of smartphones and the availability of high-speed mobile data networks, have led to the emergence of the connected vehicle. Some modern cars offer online services such as OTA SW updates, real-time traffic information, and remote diagnostics. Users also demand connectivity to their smartphones, allowing them to make and receive calls in the car, interact with GPS systems, read instant messages, and communicate with the infotainment system via voice commands using a voice assistant. Users also want to receive notifications on their phones about the status of the vehicle, whether to remind them when the car needs to be refueled or recharged, or when the next service is due. These applications can also collect data on schedules, routes, driving habits and even the installed updates [31]. As for the car itself, in some cases it may record user information for configuration purposes, such as customized views on the dashboard or preset settings for the position of seats and mirrors. This has made vehicles more vulnerable to cyberattacks by providing more entry points for malicious actors to exploit [31]. In addition, the emphasis on adding new features and connected functions has sometimes come at the expense of security, as it has been considered a secondary concern.

Electric Vehicles (EVs) also pose cybersecurity challenges, both in the car and in the power grid. For example, new home charging stations include features such as remote control of charging methods, which can be convenient but also make these devices more vulnerable [30]. In addition, Kaspersky cybersecurity researchers identified vulnerabilities in an EV charging station that could allow an attacker to damage the home power grid. The security threats ranged from stopping the vehicle charging process to setting the station to maximum current flow, which could lead to a fire [32].

As technology evolves, so does the risk of cyberattacks. With the rise of connected vehicles, the risk of cybersecurity threats is a growing concern. In the future, as today's modern, connected cars age, they may become more vulnerable. This is because the vehicles' SW and security systems are likely to become outdated and will no longer receive updates or patches from the manufacturer. As a result, the car may be more susceptible to hacking and data breaches, which can jeopardize the privacy and safety of its occupants and other road users. In addition, the car's HW components may become less reliable and secure as they age, increasing the risk of malfunctions and physical threats to the car's systems. Suppliers and OEMs must work together to address this issue and commit to designing and manufacturing safe, secure, and reliable systems.

V. Conclusions and Future Work

Cybersecurity is increasingly present in all areas. Every day, a lot of digital data is generated and spread through various channels. Sometimes this data contains sensitive information that needs to be protected.

Automotive is not different. Cybersecurity has also reached this industry and is set to be a growing trend. There are currently a lot of electronic systems connected to the vehicle's ECUs. Different communication protocols coexist inside the car for every need, both wired and wireless. All these communications must be secure and so must the in-vehicle devices.

Connectivity has made it possible for smartphones to be linked to the vehicle. Similarly, the car carries other wireless communication systems, such as GPS, 4G/5G mobile communications (to receive OTA updates or make emergency calls) or radio signals (to detect the car key or transmit tire pressure). For the future, there is talk of Drive-by-Wire, ADAS, autonomous driving and, with smart cities, Vehicle-to-Everything (V2X). This makes it necessary to secure the vehicle.

The question that remains is what will happen to the connected vehicle in the future. As a current vehicle ages, security vulnerabilities may emerge. When they are identified, it should be the responsibility of the manufacturer or supplier to provide support and fix them with an update. This will be difficult in many cases, especially if these parts are no longer manufactured and support is withdrawn. The digitization of the automotive sector will bring with it numerous cybersecurity challenges.

Acknowledgment

This work has been partially supported by the grant 'Ayudas para el desarrollo de proyectos de I+D mediante la contratación de personas doctoradas y la realización de doctorados industriales, programa BIKAINTEK 2019', by the Department of Economic Development, Sustainability and Environment of the Basque Government.

References

[1] C. Hammerschmidt, "Innovation in the car: 90% comes from electronics and software," eeNews Europe, Apr. 29, 2014. https://www.eenewseurope.com/news/innovation-car-90-comes-electronics-and-software (accessed Feb. 09, 2023).

[2] P. Mohankumar, J. Ajayan, R. Yasodharan, P. Devendran, and R. Sambasivam, "A review of micromachined sensors for automotive applications," Measurement: Journal of the International Measurement Confederation, vol. 140, pp. 305–322, 2019, doi: 10.1016/j.measurement.2019.03.064.

[3] Statista, "Car costs - Automotive electronics costs worldwide 2030," 2019. https://www.statista.com/statistics/277931/automotive-electronics-cost-as-a-share-of-total-car-cost-worldwide/ (accessed Jan. 25, 2023).

[4] X. Ferràs, E. Tarrats, and N. Arimany, "Disruption in the automotive industry: A Cambrian moment," Business Horizons, vol. 60, no. 6, pp. 855–863, 2017, doi: 10.1016/j.bushor.2017.07.011.

[5] A. Simonazzi, J. Jorge Carreto Sanginés, and M. Russo, "The future of the automotive industry: dangerous challenges or new life for a saturated market?" Institute for New Economic Thinking Working Paper Series, pp. 1–34, Nov. 2020, doi: 10.36687/inetwp141.

[6] O. Burkacky, J. Eichmann, M. Kellner, P. Keuntje, and J. Werra, "Rewiring car electronics and software architecture for the Roaring 2020s," McKinsey Center for Future Mobility, no. August, 2021.

[7] J. Á. Gumiel, J. Mabe, J. Jiménez, and J. Barruetabeña, "Introducing the Electronic Knowledge Framework into the Traditional Automotive Suppliers' Industry: From Mechanical Engineering to Mechatronics," Businesses 2022, Vol. 2, no. 2, pp. 273–290, Jun. 2022, doi: 10.3390/BUSINESSES2020018.

[8] European Automobile Manufacturers' Association (ACEA), "The Automotive Regulatory Guide." 2021.

[9] ISO/SAE, "ISO/SAE 21434:2021. Road vehicles — Cybersecurity engineering." 2021.

[10] R. Hegde, S. Kumar, and K. S. Gurumurthy, "The Impact of Network Topologies on the Performance of the In-Vehicle Network," International Journal of Computer Theory and Engineering, no. January, pp. 405–409, 2013, doi: 10.7763/ijcte.2013.v5.719.

[11] S. Kim and R. Shrestha, Automotive Cyber Security: Introduction, Challenges, and Standardization. Singapore: Springer Singapore, 2020.

[12] A. Martínez, K. A. Ramírez, C. Feregrino, and A. Morales, "Security on in-vehicle communication protocols: Issues, challenges, and future research directions," Computer Communications, vol. 180, no. September, pp. 1–20, 2021, doi: 10.1016/j.comcom.2021.08.027.

[13] G. Costantino, M. De Vincenzi, and I. Matteucci, "In-Depth Exploration of ISO/SAE 21434 and Its Correlations with Existing Standards," IEEE Communications Standards Magazine, vol. 6, no. 1, pp. 84–92, 2022, doi: 10.1109/MCOMSTD.0001.2100080.

[14] C. Schmittner, G. Griessnig, and Z. Ma, "Status of the Development of ISO/SAE 21434," in EuroSPI 2018: Systems, Software and Services Process Improvement, X. Larrucea, I. Santamaría, R. V. O'Connor, and R. Messnarz, Eds. Bilbao: Springer, 2018, pp. 504–513.

[15] K. Felser and M. Wynn, "Digitalization and Evolving IT Sourcing Strategies in the German Automotive Industry," International Journal on Advances in Intelligent Systems, vol. 13, no. 3 & 4, pp. 212–225, 2020.

[16] A. G. Marino, F. Fons, and J. M. M. Arostegui, "The Future Roadmap of In-Vehicle Network Processing: a HW-centric (R-)evolution," IEEE Access, vol. 0, no. July, pp. 69223–69249, 2022, doi: 10.1109/ACCESS.2022.3186708.

[17] F. Luo, X. Zhang, Z. Yang, Y. Jiang, J. Wang, M. Wu et al., "Cybersecurity Testing for Automotive Domain: A Survey," Sensors, vol. 22, no. 23, 2022, doi: 10.3390/s22239211.

[18] C. Bordonali, S. Ferraresi, and W. Richter, "Shifting gear s in cyber security for connected cars," McKinsey & Company, 2017.

[19] T. Królikowski and A. Ubowska, "TISAX - optimization of IT risk management in the automotive industry," Procedia Computer Science, vol. 192, pp. 4259–4268, 2021, doi: 10.1016/j.procs.2021.09.202.

[20] D. Cappelli, A. Moore, and R. Trzeciak, "Insider Theft of Intellectual Property," in The CERT Guide to Insider Threats: How to Prevent, Detect, and Respond to Information Technology Crimes (Theft, Sabotage, Fraud), Pearson Education, 2012, pp. 61–98.

[21] N. Mhaskar, M. Alabbad, and R. Khedri, "A Formal Approach to Network Segmentation," Computers and Security, vol. 103, p. 102162, 2021, doi: 10.1016/j.cose.2020.102162.

[22] S. N. Nikoi, C. Adu-Boahene, and A. Nsiah-Konandu, "Enhancing the Design of a Secured Campus Network using Demilitarized Zone and Honeypot at Uew- kumasi Campus," Asian Journal of Research in Computer Science, pp. 14–28, Jan. 2022, doi: 10.9734/ajrcos/2022/v13i130304.

[23] P. J. Ezra, S. Misra, A. Agrawal, J. Oluranti, R. Maskeliunas, and R. Damasevicius, "Secured Communication Using Virtual Private Network (VPN)," in Cyber Security and Digital Forensics - Proceedings of ICCSDF 2021, K. Khanna, V. V. Estrela, and J. J. P. C. Rodrigues, Eds. Springer Singapore, 2022, pp. 309–319.

[24] M. Fadzil, A. Kadir, M. Afif, D. Azmi, A. Nazari, and M. Rose, "Secure Communication over Virtual Private Network," vol. 35, no. 10, pp. 2129–2132, 2017, doi: 10.5829/idosi.wasj.2017.2129.2132.

[25] E. Babetto, "Blockchain ed Economia Circolare: Tracciabilità e Sostenibilità," Università di Padova, 2022.

[26] S. Povolny, "Model Hacking ADAS to Pave Safer Roads for Autonomous Vehicles," McAfee Blog, 2020. https://www.mcafee.com/blogs/other-blogs/mcafee-labs/model-hacking-adas-to-pave-safer-roads-for-autonomous-vehicles/ (accessed Nov. 18, 2022).

[27] B. Nassi, Y. Mirsky, D. Nassi, R. Ben-Netanel, O. Drokin, and Y. Elovici, "Phantom of the ADAS: Securing Advanced Driver-Assistance Systems from Split-Second Phantom Attacks," in Proceedings of the 2020 ACM SIGSAC Conference on Computer and Communications Security, Oct. 2020, no. Report 2020/085, pp. 293–308, doi: 10.1145/3372297.3423359.

[28] European Automobile Manufacturers' Association (ACEA), "Average age of the EU vehicle fleet, by country." 2022.

[29] Statista, "Size of the global connected car fleet in 2021, with a forecast for 2025, 2030, and 2035, by region," 2021.

[30] Z. Pourmirza and S. Walker, "Electric Vehicle Charging Station: Cyber Security Challenges and Perspective," 2021 9th IEEE International Conference on Smart Energy Grid Engineering, SEGE 2021, pp. 111–116, 2021, doi: 10.1109/SEGE52446.2021.9535052.

[31] D. Colombo, "How I got access to 25+ Tesla's around the world. By accident. And curiosity.," Medium, 2022. https://medium.com/@david_colombo/how-i-got-access-to-25-teslas-around-the-world-by-accident-and-curiosity-8b9ef040a028 (accessed Jan. 25, 2023).

[32] D. Sklyar, "ChargePoint Home security research," Kaspersky Lab Security Services, 2018.

International Academy, Research,
and Industry Association (IARIA)
412 Derby Way
Wilmington, DE 19810

ISBN 978-1-7138-7129-3

CURRAN ASSOCIATES INC.
proceedings.com

Twelfth International Conferences on Advances in Multimedia (MMEDIA 2020)

Held at NexComm 2020

Lisbon, Portugal
23 - 27 February 2020

IARIA